D0204317

Milestones in Science and Technology:

The Ready Reference Guide to Discoveries, Inventions, and Facts

Milestones in Science and Technology:

The Ready Reference Guide to Discoveries, Inventions, and Facts

by Ellis Mount and Barbara A. List

The rare Arabian Oryx is believed to have inspired the myth of the unicorn. This desert antelope became virtually extinct in the early 1960s. At that time several groups of international conservationists arranged to have 9 animals sent to the Phoenix Zoo to be the nucleus of a captive breeding herd. Today the Oryx population is over 400, and herds have been returned to reserves in Israel, Jordan, and Oman.

The Oryx Press
2214 North Central at Encanto
Phoenix, AZ 85004-1483

Published simultaneously in Canada

Printed and Bound in the United States of America

The paper used in this publication meets the minimum requirements of American National Standard for Information Science—Permanence of Paper for Printed Library Materials, ANSI Z39.48, 1984.

Library of Congress Cataloging-in-Publication Data

Mount, Ellis.
Milestones in science and technology.

Bibliography: p.
Includes indexes.
1. Science—Handbooks, manuals, etc. 2. Inventions—Handbooks, manuals, etc. I. List, Barbara A. II. Title.
Q199.M68 1987 509 87-12352
ISBN 0-89774-260-5 (alk. paper)

Contents

Preface

The disciplines of science and engineering (or technology) represent complex patterns of events, some of great importance to the world, but many of lesser importance. Because of the large number of events or discoveries, the purpose of this book is to provide information on a selection of what the compilers feel are the 1000 most significant topics in all areas of science and technology. Obviously no two compilers would make the same selections. We tried to strike a balance among the many fields in which science and technology are significant. Regarding time periods, our choices range from the prehistoric era up to the present year. Topics cover basic discoveries as well as practical inventions, and range from relativity and genetics to the electric motor and DDT.

The main body of the book is arranged alphabetically by the name of the invention or the discovery. Each entry gives a brief description of the topic as well as the dates these events occurred, the names of the people involved, and the nationalities of the persons named. In addition each topic is assigned a broad category, such as physics or medicine. Cross-references to any related entries are provided at the end of the descriptions, and any entries mentioned within the descriptions are set in boldface type.

There are four indexes for the book, arranged, respectively, by year of the event, by names of people involved, by nationalities of the people, and by names of broad categories. Each of these four indexes refers to the name of the invention or discovery, thus making it possible to determine what happened in a particular year, or what inventions are connected to a particular person, or what inventions were made by people of a particular nationality, or what discoveries involved a given broad category.

This book is designed for use by several audiences, such as students, laypersons, reference librarians, and practitioners working outside their own fields of expertise. It might well prove to be helpful to historians wanting to discover relationships for a particular time period or in a particular country of interest to them.

Each entry contains the name of an additional book or reference source that might be of interest to readers desiring more information on a particular topic. We found, however, that it was rarely possible to cite any one source that fully described the topic in detail or that gave all the dates and factual data involved. As compilers we thus had to use our judgment as to which of several possible sources to cite since we did not list more than one additional title for each topic. Although an effort was made to refer readers to books and other sources that were known to be in print, in some cases it was necessary to cite works no longer in print but nevertheless apt to be found in most medium-sized public libraries.

Those who have attempted to determine accurately the dates of events and the names of the people involved have undoubtedly discovered, as did we, that there are many disagreements among even reputable reference sources about such details as dates and names. Again, in many instances we had no choice but to use our best judgment as to which authority to quote.

We made use of the 100 or so basic categories which are used in the third edition of the *McGraw-Hill Dictionary of Scientific and Technical Terms* (1984) for identifying broad disciplines, such as astronomy or biochemistry.

In a few instances we had to add categories to those found in this reference source, but essentially we found them to be quite satisfactory. Not all of them were used because we decided at the outset that certain fields would not be included in the book. These are anthropology, geography, architecture, paleontology, and parapsychology. Diseases are included only insofar as they relate to significant developments, such as Edward Jenner's discovery of the smallpox vaccine.

We have provided an appendix that gives full citations for works cited in the entries as additional sources to consult; in addition we listed a few titles which proved to be useful in our work, even if no particular citation was made to such reference tools.

There are many excellent sources available that deal with the development of discoveries and inventions, but we feel that our book differs sufficiently from these other titles to justify the addition of one more book in this field.

We would like to acknowledge the strong support and assistance given to us by our respective families, without which this book could not have been prepared. Particular thanks are due to Katherine Mount for inputting the bulk of the book into a computer and to Marsha Mount for her expertise in computer operations. Also many of our friends and colleagues at our places of work were most helpful. Special thanks are due to the two people who reviewed our entries in their respective fields of interest, namely Mary Rounsifer and Robert G. Krupp: they both made many excellent suggestions for improving the book.

We would also like to acknowlege the help we received from the staffs and the collections at the following libraries: University of Michigan Libraries, Columbia University Libraries, Ann Arbor (MI) Public Library, Teaneck (NJ) Public Library, and Bergenfield (NJ) Public Library.

Alphabetical Listing of Terms

A

Abacus — MATHEMATICS

The abacus was an ancient device for making calculations by moving counters or beads strung on wires. The device is known to have been used in both ancient Greece and Rome; Herodotus, the Greek historian, mentioned use of the abacus around 400 B.C. in Egypt. Ancient types of abacus have also been found in China, thought by some to have been taken there by Arabic traders. The first known use of them in China was during the twelfth century, where the device was called the *suan-pan.* They were reinstated in Europe around 1000 A.D. by Pope Sylvester II, but were gradually replaced by Arabic notation beginning around 1200, although the notation was not widely used by the general population for centuries.

Additional reading: Asimov. *Asimov's new guide to science;* p. 855–856.

Abrasive — MATERIALS

Abrasives as a class are simply hard substances used to rub material from softer substances. It is believed that sand was used for polishing weapons made of stone as far back as 25,000 B.C. Diamonds were used as abrasives in India in 700 B.C. Other early abrasive materials were garnet, pumice, and emery. By the Middle Ages grinding wheels of quartz and flint fragments were used. Sandpaper was developed later, then came fine-grained materials such as emery paper. The first important synthetic abrasive was **carborundum,** made in 1891. **Synthetic diamonds** were first made in 1955.

Additional reading: *Illustrated science and invention encyclopedia;* v. 1, p. 6–7.

Absolute Zero. *See* Kelvin Temperature Scale; Liquefaction of Gas.

Accelerometer — MECHANICAL ENGINEERING

Devices for the measurement of acceleration are of two types—those for linear (straight line) acceleration and those for angular (twisting) acceleration. The first accelerometer was built in 1783 by George Atwood, an English physicist.

Additional reading: *Illustrated science and invention encyclopedia;* v. 1, p. 8.

Acetic Acid — CHEMISTRY

One of the earliest types of **acid** to be discovered; it was made around 800 A.D. by the Arabian alchemist Geber, who distilled vinegar to produce a strong acetic acid. Over a thousand years later the German chemist Adolf Kolbe became the first to synthesize organic compounds from inorganic materials, making acetic acid in 1845. This disproved the belief that organic compounds could not be formed from inorganic materials.

Additional reading: Asimov. *Asimov's biographical encyclopedia;* p. 48–49, 397.

Acetylcholine — BIOCHEMISTRY

In 1921 German-born physiologist Otto Loewi (U.S.) discovered that a chemical substance was involved in nerve impulses. Loewi called this substance "Vagusstoff" and found that it stimulated organs and was released from the vagus nerve. Sir Henry Hallett Dale, an English biologist, was also working in this area in the 1920s. He isolated the same substance, now called "acetylcholine," and showed that it affected organs in the same way that nerves belonging to the parasympathetic system did. These advances ushered in the study of neurochemistry and introduced the awareness that the nervous system was ruled by more than electric impulses. Loewi and Dale shared the 1936 Nobel Prize in medicine and physiology for their work.

Additional reading: Sourkes. *Nobel prize winners in medicine and physiology 1901–1965;* p. 191–200

Acetylsalicylic Acid — PHARMACOLOGY

In 1859 German chemist Adolph Wilhelm Hermann Kolbe discovered the "Kolbe" reaction when he discovered that he could get double acids by applying electrolysis to organic compounds. His finding made it possible to prepare large quantities of salicylic acid. After the German chemical firm Beyer introduced acetylsalicylic acid under the trade name Aspirin in 1899, it was possible to produce the new drug inexpensively using the Kolbe reaction.

Additional reading: Arnow. *Health in a bottle;* p. 61–64.

Acid — CHEMISTRY

Many compounds can be classed as acids; they all have the characteristic that when an acid is dissolved in water its hydrogen atoms split away from the rest of its molecules and become electrically charged ions with a strong tendency to react with other materials. This definition was proposed separately by the Swedish chemist Svante Arrhenius and the German chemist Friedrich W. Ostwald in 1884. A newer definition proposed in 1923, by the Danish chemist J. N. Brenstad, is that an acid is a material which can act as a source of protons, as distinct from a **base,** which can accept protons. This broadened the definition of acid and includes solvents other than water. (*See also* names of specific types of acids)

Additional reading: Asimov. *Asimov's biographical encyclopedia;* p. 544–545, 577–579, 670–671.

Acoustics. *See* Architectural Acoustics.

Acquired Immunological Tolerance. *See* Immunological Tolerance.

Adding Machine. *See* Calculating Machine.

Adenine. *See* Nucleic Acid.

Adhesive — MATERIALS

Natural adhesives were used in Egypt as far back as 1300 B.C.; flour paste, bitumen, egg white, and beeswax were examples of such substances. In the nineteenth century rubber and pyroxylin cements were developed, but in the twentieth century emphasis was on synthetic adhesives or synthetic resins. Either thermoplastic (melting when heated) or thermosetting (hardening when heated), adhesives are used for many purposes and come in many forms—liquids, powders, and tapes. One of the earliest tapes was invented by the German pharmacist Paul Beiersdorf, who devised a plastic-covered bandage in 1882. In 1925 Richard Drew (U.S.) developed at the 3M Company a special tape used for masking an item being painted. By 1930 cellular adhesive tape (called "Scotch tape") was available.

Additional reading: *Illustrated science and invention encyclopedia;* v. 1, p. 16–17.

Adrenaline. *See* **Epinephrine.**

Aerial Photography GRAPHIC ARTS

The first aerial photograph was taken from a balloon over France in 1856. In 1862 a photographer in a tethered balloon took photographs of Confederate troop positions in Virginia. By 1909 photos had been taken from airplanes, which continued during World War I, particularly by the British photographer Lord Brabazon, who developed cameras mounted in the floor of the plane. In World War II special photographic reconnaissance planes were used. After the development of **satellites,** greater coverage and more detailed mapping became possible, including **infrared photography.**

Additional reading: *Illustrated science and invention encyclopedia;* v. 1, p. 20–23.

Aerodynamics. *See* **Fluid Mechanics.**

Aerosol Spray MECHANICAL ENGINEERING

In 1926 the Norwegian inventor Eric Rotheim devised the aerosol spray when he discovered that a product could be sprayed by putting a gas or a liquid into a container to build internal pressure. In 1939 a patent was granted to Julian S. Kahn (U.S.) for a disposable spray can but it was not developed much until 1941. That same year a U.S. patent was won by L. D. Goodhue (U.S.) for a type of refillable aerosol spray. However, this did not come into common use until the 1950s.

Additional reading: *Illustrated science and invention encyclopedia;* v. 1, p. 28.

Agglutination Reaction IMMUNOLOGY

In 1896 English bacteriologist Herbert Edward Durham and German bacteriologist Max von Gruber discovered specific agglutination. In the same year and based on Durham's and von Gruber's work, the agglutination reaction was discovered by French physician Fernand Isidor Widal when he found that the blood serum taken from a typhoid fever patient caused a culture of typhoid bacilli to lose motility and to agglutinate or clump together. Widal referred to this as a "serodiagnostic" technique and thereby introduced the first "serum" diagnosis.

Additional reading: Reiser. *Medicine and the rise of technology;* p. 136.

Agglutinogen. *See* **Blood Groups.**

Air Composition CHEMISTRY

The composition of ordinary air puzzled chemists for centuries, but in 1778 the French chemist Antoine Lavoisier was the first scientist to state that air consisted essentially of a mixture of oxygen and nitrogen, a conclusion that other scientists had merely suspected to be true.

Additional reading: Asimov. *Asimov's biographical encyclopedia;* p. 222–226.

Air Compressor MECHANICAL ENGINEERING

The earliest means of compressing air was by foot-operated blow sacks used by the Egyptians around 1500 B.C. to increase the air flow on furnaces for smelting metals. For many centuries manual exertion was involved, such as the ordinary tire pump. In 1799 George Medhurst (England) compressed air for a motor in a mine, while in 1802 William Murdock (England) used compressed air for driving lathes and operating an elevator. In 1830 Thomas Cochrane (England) used compressed air for tunnelling, and a compressed air drill for use on rocks was invented in 1857 by the French engineer Germaine Sommeiler. Modern types are either reciprocating compressors (in which a piston slides back and forth) or rotary compressors (in which there are revolving blades) .

Additional reading: *Illustrated science and invention encyclopedia;* v. 5, p. 628–629.

Air Conditioner MECHANICAL ENGINEERING

Following a study of air humidity in 1902, William Carrier (U.S.) in 1904 devised an air cleaner which included water spraying. In 1911 he invented a complete air conditioning system. In 1919 the first air-conditioned movie theater opened, as did the first department store so treated, both in the United States. Individual room air conditioners were designed by H. H. Schutz (U.S.) and J. Q. Sherman (U.S.), probably in the 1930s. By the 1960s millions of room air conditioners were in use. Air conditioners for automobiles were developed in the 1950s.

Additional reading: *Illustrated science and invention encyclopedia;* v. 1, p. 34–35.

Air Cushion Vehicle. *See* **Hovercraft.**

Air Pollution. *See* **Pollution Control.**

Aircraft AEROSPACE ENGINEERING

Efforts to achieve powered flight made use of developments stemming from the invention of **gliders** and **airships.** Early airplane designs were powered by rubber bands, by flapping of wings, or even by a heavy steam engine. Finally in 1903 the Americans Wilbur and Orville Wright made a free flight of 120 feet in a gasoline-powered airplane. In 1905 they could stay aloft for half an hour. By 1927 equipment had reached the stage that Charles A. Lindbergh (U.S.) could fly a single-engine plane on a solo flight from New York to Paris. By 1939 German and English engineers were designing fighter planes with **jet engines.** Efforts to achieve **supersonic flight** succeeded in 1947 when a U.S. plane flew faster than the speed of sound.

Additional reading: Clark. *Works of man;* p. 240–260.

Aircraft Carrier. *See* **Catapult.**

Aircraft Engine AEROSPACE ENGINEERING

One of the first engines developed specially for aircraft was built in 1903 by Orville and Wilbur Wright (U.S.), a four-cylinder model producing 12 horsepower. A radical new design appeared in 1907, a light-weight rotary engine. Still another design was the air-cooled radial engine, made in England in 1918. The air-cooled Rolls-Royce Merlin engine (England) powered most of the British planes in World War II. **Jet engines** took over the market after World War II.

Additional reading: *Illustrated science and invention encyclopedia;* v. 1, p. 23–27.

Aircraft Instrument AEROSPACE ENGINEERING

The various instruments commonly used to fly modern aircraft had their origins in a number of early inventions. For example, the rate of climb indicator depends upon the measurement of atmospheric pressure, the principle on which the barometer was based by its inventor, the Italian physicist Evangelista Torricelli, in 1644. Both the artificial horizon and the rate of turn indicator depend upon gyroscopes, first studied scientifically by the French physicist Leon Foucault in 1852. The altimeter is based on radar, developed by the Scottish engineer Sir Robert Watson-Watt in 1935. The automatic pilot depends upon servomechanisms, which date back to the governor for a steam engine invented by the Scottish engineer James Watt in 1769. Many other examples could be cited.

Additional reading: *Illustrated science and invention encyclopedia;* v. 10, p. 1268–1271.

Airship AEROSPACE ENGINEERING

As progress was made in the development of **balloons,** inventors began to devise ways of directing the flight of balloons. The first successful flight of an airship took place in 1852 in a trip of 17 miles in a craft built by the French inventor Henri Giffard. The first successful flight of a freely guided airship took place in 1884 in a motor-powered airship designed by the French inventors A. C. Krebs and Charles Renard. The world's first rigid airship was built by David Schwarz (Austria) in 1897. Inspired by Schwarz's work, Count Ferdinand von Zeppelin (Germany) also launched a rigid airship (known as a dirigible) in 1900; by 1910 to 1914 he was operating the world's first airline. Crashes in storms of several airships later ended their public use.

Additional reading: Feldman. *Scientists & inventors;* p. 194–195.

Alchemy CHEMISTRY

Alchemy was practiced as far back as 100 A.D. in Alexandria. Known today as a pseudo-science centered primarily on changing base metals into gold, alchemy has nevertheless made some contributions to modern science. For example, many pieces of laboratory equipment, such as stills, beakers, and condensers were used by alchemists, as were certain techniques like sublimation, crystallization, or distillation. As modern chemistry developed, these skills and equipment were ready to be used by scientists.

Additional reading: *Illustrated science and invention encyclopedia;* v. 1, p. 61–64.

Algebra MATHEMATICS

An ancient papyrus written by the Egyptian priest Ahmes around 1700 B.C. shows that he was familiar with problems in arithmetic, algebra, and geometry. He used symbols to stand for unknown numbers and for our "=" sign. Around 600 B.C. the Greek astronomer Thales studied with Egyptian priests and later taught his avid student, the Greek scholar Pythagoras. The latter not only excelled in geometry but also made advances in algebra, such as his theory of irrational numbers. Important progress in algebra was made by the Greek mathematician Diophantus around 250 A.D. He solved simultaneous and quadratic equations, and equations of higher order. He also used symbols for unknown quantities. Algebra apparently did not reach Europe until the fifteenth century.

Additional reading: Newman. *World of mathematics;* v. 1, p. 12, 79, 81–82, 113–117.

Alloy METALLURGY

Mixtures involving one or more metals are thousands of years old. The first alloy commonly used, bronze (a mixture of tin and copper), is known to have existed around 3000 B.C. By definition, an alloy must have metallic properties and consist of two or more elements (usually metallic elements). Alloys are indispensible to our society; thousands of them have been developed.

Additional reading: *Illustrated science and invention encyclopedia;* v. 1, p. 73–76.

Alpha Particle NUCLEAR PHYSICS

The discovery of **radioactivity** in 1896 led the British physicist Ernest Rutherford to study the emissions from radioactive substances. He found both electrons and a less penetrating emission, which in 1897 he called beta particles and alpha particles, respectively. He learned that the alpha particles had a positive charge. Later they were seen to be identical to the nucleii of helium atoms, consisting of two **neutrons** and two **protons**.

Additional reading: Asimov. *Asimov's new guide to science;* p. 317–319.

Alternating Current Generator. *See* **Electric Generator.**

Alternating Current Motor. *See* **Electric Motor.**

Alternator. *See* **Electric Generator.**

Altimeter AEROSPACE ENGINEERING

The decrease in air pressure as altitude increases is the basis for the ordinary altimeter for airplanes. By converting the air pressure to altitude, the instrument indicates height above sea level. Over high terrain the readings would not indicate height above the ground. The solution to this is use of a radar altimeter. One of the earliest uses was in 1783 when the French physicist Jacques Charles sent a barometer up with a balloon to measure altitude. (*See also* Radar)

Additional reading: *McGraw-Hill encyclopedia of science and technology;* v. 1, p. 387–390.

Aluminum CHEMISTRY

Aluminum, atomic number 13, was first isolated in 1825 by the Danish chemist Hans Christian Oersted. It does not exist freely in nature, requiring complicated extraction methods. In 1845 the German physicist Friedrich Wohler succeeded in purifying a small quantity. In 1886 the American chemist Charles Hall and the French chemist Paul Heroult independently discovered an economical, electrolytic method of extraction of pure aluminum It is used as an alloy, in the transportation equipment industry and in medical applications. Chemical symbol is Al.

Additional reading: *Van Nostrand's scientific encyclopedia;* p. 104–107.

AM Radio. *See* **Amplitude Modulation.**

Americum CHEMISTRY

A radioactive metal of the actinide series, americum, atomic number 95, is also one of the transuranium elements. It was discovered in 1954 by the American scientists Glenn T. Seaborg, A. Ghiorso, R. A. James, and L. O. Morgan. It is produced by bombarding heavy elements like uranium or plutonium. Chemical symbol is Am.

Additional reading: *Van Nostrand's scientific encyclopedia;* p. 115–116.

Amino Acid BIOCHEMISTRY

In 1806 Louis Nicolas Vauquelin, a French chemist, isolated asparagine, a compound in asparagus that turned out to be the first amino acid occurring in **protein** to be discovered. English chemist and physicist William Hyde Wollaston discovered the second amino acid, cystine, in 1810 in a bladder stone. Eventually more than 80 amino acids were discovered, with 20 of those serving as basic building blocks of protein.

Additional reading: Asimov. *Asimov's new guide to science;* p. 551–554.

Ammeter ELECTRICITY

Ammeters are instruments for measuring electric currents. They have an electromagnet and a permanent magnet, either one of which is movable, although the moving coil type is more uneven. They are not as sensitive as **galvanometers** due to their design, which causes only a small proportion of the current to flow through the coil. The first one was invented in 1820 by Hans Christian Oersted (Denmark).

Additional reading: *Illustrated science and invention encyclopedia;* v. 1, p. 89–90.

Amphibious Aircraft. *See* **Seaplane.**

Amphibious Vehicle MECHANICAL ENGINEERING

Vehicles designed to travel over land and through water are called amphibious vehicles. Both personal and military uses are common. The first known example was built in 1805 by the American inventor Oliver Evans; it was designed to be a steam dredger. Military uses have ranged from small Jeep-size amphibious vehicles up to 38-ton landing craft. A related type is the swamp buggy, which has large truck or tractor type tires that drive the vehicle through thick mud or water. (*See also* Hovercraft)

Additional reading: *Illustrated science and invention encyclopedia;* v. 1, p. 96–100.

Amplitude Modulation
COMMUNICATIONS

The technique of modifying the amplitude of a constant frequency carrier wave is called amplitude modulation. This was the first system used for commercial radio broadcasting beginning in 1921, and no doubt is still the more common type compared to stations using **frequency modulation.** Amplitude modulation is more subject to static but can normally broadcast for longer distances than frequency modulation. One of the inventions that simplified the use of AM radios was the superheterodyne receiver devised by the American engineer Edwin H. Armstrong around 1916.

Additional reading: *Illustrated science and invention encyclopedia;* v. 1, p. 103.

Amytal. *See* Barbiturate.

Analog Computer
COMPUTER SCIENCE

An analog computer, which makes use of mechanical or electrical devices to represent numbers being manipulated, was designed by the Scottish physicist Lord Kelvin around 1880. Although it could not be built, it was said to be the inspiration for the American engineer Vannevar Bush, who succeeded in 1930 in building the first analog computer. He called it a differential analyzer. It was used to aim antiaircraft guns. Later models have been used for many special purposes, such as flight simulators for training pilots.

Additional reading: Ritchie. *The computer pioneers;* p. 21–25.

Analog Converter. *See* Digitizer.

Analytic Geometry
MATHEMATICS

The first application of algebra or analytical methods to geometry was primarily the work of Rene Descartes (France) around 1637. His aim was to provide a geometrical interpretation of algebra as well as the geometrical construction of solutions of certain algebraic equations. He was followed by Jan de Witt (Holland) who contributed work on conic sections around 1659. Analytic geometry of higher dimensions was studied by Arthur Cayley (England) around 1843 and Herman Grassman (Poland) also around 1843.

Additional reading: Bynum. *Dictionary of the history of science;* p. 15–16.

Anaphylaxis
MEDICINE

In 1902 Charles Robert Richet, a French physiologist, discovered that the second dose of an antigen may sometimes kill an organism by inducing fatal shock. When the antigen was first injected, the animal produced an **antibody** which caused a deadly reaction with the subsequent injection. Richet named this phenomenon "anaphylaxis" from the Greek words meaning "over protection." The 1913 Nobel Prize in medicine and physiology was awarded to Richet for his discovery.

Additional reading: Asimov. *Asimov's new guide to science;* p. 685.

Anatomy
BIOLOGY

Diocles, a Greek physician born about 350 probably wrote the first anatomy book and was the first to use the word. Herophilus, a Greek born about 320 B.C., is known as the first careful anatomist and compared human bodies with those of animals. In 1316 the Italian Mondino de'Luzzi, on the basis of his own dissections, wrote the first book entirely devoted to anatomy. Andreas Vesalius, a Flemish physician, ushered in the beginning of modern anatomy in 1543, when he published the first accurate, illustrated human anatomy book entitled *De Corporis Humani Fabrica.* Vesalius' book made accurate surgery possible for the first time. French zoologist Baron Georges Cuvier founded comparative anatomy in 1805 with his book *Lesson in Comparative Anatomy.*

Additional reading: Magner. *A history of the life sciences;* p. 50–54, 65–70, 98–106.

Android. *See* Robot.

Anemometer
METEOROLOGY

The inventor of the first known device for measuring the velocity of the wind, the anemometer, was the English astronomer Robert Hooke in 1644. The simplest type consists of three or four cups which rotate around a vertical shaft. For use in turbulent air a hot-wire anemometer is used. It measures the amount of cooling of a heater wire when exposed to moving air, calibrated to read in terms of velocities.

Additional reading: *McGraw-Hill encyclopedia of science and technology;* v. 1, p. 507–508.

Anesthesia
PHYSIOLOGY

In 1266 Theodoric Borgognoni of Lucca (Italy) advocated the use of soporific sponges soaked with narcotics for inducing sleep in surgical patients. Henry Hickman (England) administered carbon dioxide to animals prior to surgery in 1824. The earliest use of intravenous anesthesia occurred in 1872 when C. Ore (France) injected a solution of chloral hydrate into a patient. (*See also* Chloroform; Cocaine; Ether; Novocain)

Additional reading: Sneader. *Drug discovery;* p. 15–41.

Angiosperm
BOTANY

In the early 1800s Scottish botanist Robert Brown classified the higher plants, giving the name "angiosperm" to flowering plants. (*See also* Gymnosperm)

Additional reading: Morton. *History of botanical science;* p. 373–376.

Animal and Plant Classification
BIOLOGY

The Greek philosopher Aristotle, who lived between 384 B.C. and 322 B.C., developed a classification scheme for animals in which he arranged over 500 species into hierarchies. A system of binomial nomenclature still in use today was developed in 1735 by Swedish botanist Carolus Linnaeus. Linnaeus gave each plant or animal two names, the first to designate genus or related group, and the second to give species or specific name within the genus. Linnaeus popularized his system when he published "*Systema Naturae.*"

Additional reading: Magner. *A history of the life sciences;* p. 348–354.

Antenna
COMMUNICATIONS

A device for transmitting or receiving electromagnetic radiation, such as radio waves or radar signals, is called an antenna. Although in 1891 Edouard Branly (France) found that the reception of electrical signals from approaching storms was improved by erecting long pieces of metal, the first true antenna was invented by Aleksandr Popov (Russia) in 1895. After following Branly's lead, Popov built a radio set that allowed him to send a signal 250 meters. Around the same time Guglielmo Marconi (Italy) was also using antennas in his experiments. Many types of antennas for radio and radar systems have since been devised.

Additional reading: *Illustrated science and invention encyclopedia;* v. 1, p. 18–20; v. 14, p. 1910–1918.

Anthrax Vaccine
IMMUNOLOGY

French chemist Louis Pasteur developed the anthrax vaccine by culturing anthrax bacteria at a high temperature, thereby producing a weakened strain that could be used to immunize livestock. In 1881 he gave a dramatic public demonstration in which he inoculated sheep, goats, and cattle with his weakened strain. Other animals were not given the vaccination. After a number of days had passed, Pasteur inoculated all of the animals with the full-strength anthrax bacteria. Only those animals that had received the vaccine survived, and Pasteur thus demonstrated the effectiveness of his treatment.

Additional reading: Magner. *A history of the life sciences;* p. 257–261.

Antibiotic MICROBIOLOGY

Scottish bacteriologist Sir Alexander Fleming discovered the first effective antibiotic, **penicillin,** in 1928. It was not until 1941, however, that Russian-born Selman Abraham Waksman (U.S.) coined the term "antibiotic" to describe those chemicals obtained from microorganisms that killed bacteria. The first of the tetracycline antibiotics, aureomycin, was discovered and introduced in 1948 by Benjamin Minge Duggar, an American botanist. In 1950 A. C. Finlay (U.S.) discovered oxytetracycline which Pfizer marketed under the tradename Terramycin. (*See also* Streptomycin)

Additional reading: Arnow. *Health in a bottle;* p. 76–78.

Antibody IMMUNOLOGY

In 1890 German bacteriologist Emil Adolf von Behring found that the blood serum of immunized animals contained factors that killed the specific organisms used for the immunization. Those factors are today called "antibodies." Gerald Maurice Edelman, an American biochemist, determined in 1969 the **amino acid** structure of a typical antibody made up of over 1,000 amino acids. Edelman shared the 1972 Nobel Prize in medicine and physiology for his work. (*See also* Monoclonal Antibody)

Additional reading: Asimov. *Asimov's new guide to science;* p. 683–688.

Anticoagulant PHARMACOLOGY

In 1869 sodium phosphate was introduced as the first anticoagulant to prevent coagulation of blood. By World War I, sodium citrate was found to be an innocuous and effective anticoagulant, and proved to be invaluable during the process of **blood transfusion.** In 1884 J. B. Haycraft established that an anticoagulant factor was present in leeches. The effectiveness of phospholipids was demonstrated in 1922 by William Howell (U.S.). He named his anticoagulant "heparin," and in 1935 an international standard was set for the sodium salt contained therein.

Additional reading: Sneader. *Drug discovery;* p. 160–164.

Antidepressant. *See* **Iproniazid.**

Antihistamine PHARMACOLOGY

In 1937 the Swiss-born pharmacologist Daniel Bovet (France) discovered a group of compounds that countered the congestion frequently brought on by allergies. Since such symptoms were thought to be a result of the production of histamines, Bovet called his compounds "antihistamines." Bovet introduced the first antihistamine to the public in 1944 and called it "pyrilamine." The 1957 Nobel Prize in medicine and physiology was awarded to Bovet for his work with antihistamines.

Additional reading: Sourkes. *Nobel prize winners in medicine and physiology 1901–1965;* p. 342–351.

Antimatter PARTICLE PHYSICS

The acceptance of the 1930 theory of the English physicist Paul Dirac that each particle should have an antiparticle has led to questions about the existence of antimatter, that is, substances made up of antiparticles. The question appeared answered when bombardment of beryllium with protons in 1965 at the American Brookhaven Laboratories resulted in combinations of antiprotons and antineutrons. Antihelium -3 has also since been produced. However, there are serious doubts as to whether antimatter exists naturally in the universe. So far there is no evidence that it does.

Additional reading: Asimov. *Asimov's new guide to science;* p. 331–332, 342–343.

Antimony CHEMISTRY

A lustrous, brittle, blue-white metal, antimony, atomic number 51, occurs most commonly as the sulfide stibnite, located largely in China. It is often used in alloys with lead, tin, or copper. Its sulfide form was known to the ancients as a cosmetic material and as a medicine, while metallic antimony was mentioned in the writings of Dioscorides, a Greek scholar around 100 A.D., although it is believed to have been used as early as 4000 B.C. Its preparation was first described around 1604 by the German chemist Basil Valentine. Chemical symbol is Sb.

Additional reading: *Van Nostrand's scientific encyclopedia;* p. 198–199.

Antineutrino PARTICLE PHYSICS

The prediction in 1931 of the existence of the **neutrino** gave rise to the prediction of the antineutrino, in order to conform to various basic principles of nuclear structure. This was borne out in 1953 when two American physicists, Frederick Reines and Clyde Cowan, studied tanks of water set in a nuclear reactor. They were successful in detecting antineutrinos.

Additional reading: Asimov. *Asimov's new guide to science;* p. 348–351.

Antiparticle PARTICLE PHYSICS

A mathematical study of subatomic particles convinced the English physicist Paul Dirac in 1930 that each particle should have an antiparticle. Thus there should be an antielectron that was just like an electron except that it had a positive electric charge rather than a negative one. This was confirmed by the discovery in 1932 of the **positron.**

Additional reading: Asimov. *Asimov's new guide to science;* p. 331–332.

Antiproton NUCLEAR PHYSICS

The high energy sources needed to produce certain nuclear particles were provided by the invention of the bevatron in 1954. The next year the American physicists Owen Chamberlain and Emilio Segre used the bevatron to bombard copper with protons, detecting the presence of antiprotons.

Additional reading: Asimov. *Asimov's new guide to science;* p. 340–341.

Antisepsis MEDICINE

In 1847 the Hungarian physician Ignaz Philipp Semmelweiss forced doctors working under him in Viennese hospitals to wash their hands with strong chemicals before touching patients to reduce the incidence of death from childbed fever. Although Semmelweiss was defeated by resentful doctors, the **germ theory** of French chemist Louis Pasteur, developed in the 1860s, supported his ideas. In 1865 Baron Joseph Lister, an English surgeon, learned of Pasteur's research and began using carbolic acid to kill germs in surgical wounds. In doing so Lister founded antiseptic surgery. William Stewart Halstead (U.S.) used rubber gloves in 1890 because they are easily sterilized and thereby introduced aseptic surgery.

Additional reading: Ackerknecht. *A short history of medicine;* p. 186–188, 191.

Antitoxin. *See* **Diptheria Antitoxin; Tetanus Vaccine.**

Appleton Layers. *See* **Ionosphere.**

Aqualung OCEANOGRAPHY

The need for allowing extended underwater exploration for divers brought about the development in the 1930s of self-contained underwater breathing apparatus by a British firm. It was found that air needed to be pressurized to match the depth of the diver, and this was met with the invention in 1943 of the aqualung, developed by the French explorer Jacques-Yves Cousteau and the French engineer Emile Gagnan. It consisted of a breathing device that contained a special valve that would furnish pressurized air when the diver breathed in and cut off air when breathing out. Cousteau reached a depth of 60 feet on his first test, later extended to over 200 feet. (*See also* Bathyscaphe)

Additional reading: Feldman. *Scientists & inventors;* p. 304–305.

Aqueduct
CIVIL ENGINEERING

The construction of aqueducts for bringing water supplies to cities dates back to around 700 B.C., when Sennacherib (Assyria) built a stone canal which crossed a wide valley in an aqueduct that was coated with bitumen so as to be waterproof. The Greek engineer Eupalinus of Megara built a long underground aqueduct around 500 B.C. The Romans depended heavily on aqueducts; Rome used 280 miles of them, providing 38 gallons of water per person. An aqueduct near Nimes (France), built around 100 A.D., is 900 feet long and has a maximum height of 160 feet, still carrying water. In later times large aqueducts were still being built. In 1847 an aqueduct 51 miles long was built near Marseilles, France; it crossed a river at a height of nearly 300 feet on a span 1,300 feet long.

Additional reading: Derry. *Short history of technology;* p. 53, 160, 163, 170, 423.

Arabic Numerals
MATHEMATICS

For centuries mathematical calculations were done using cumbersome systems for notation, particularly those developed by the Egyptians, the Greeks, and the Romans. Around 500 A.D. the Hindu astronomer Aryabhata devised a system in which the actual numerical value of a symbol would depend on its position, such as the symbol for 3 stands for 3 in the first position, and for 30 in the second position, etc. At about this time the symbol for zero was accepted. When the Arabs conquered India after 700 A.D., they were impressed by the Hindu system and imported Indian scholars into Baghdad to teach Arabs the system. Not until 1202 was it adopted in Europe, thanks to the work of Leonardo Fibonacci (Italy), but not until around 1500 was its use reasonably complete.

Additional reading: *Illustrated science and invention encyclopedia;* v. 21, Section 33.

Architectural Acoustics
ACOUSTICS

The process of designing concert halls and auditoriums so that sound can be heard clearly was advanced by the criteria first set forth by the American physicist Wallace C. Sabine around 1896. His principles, now known as Sabine's law, have invariably contributed to the construction of auditoriums with excellent acoustics.

Additional reading: Feldman. *Scientists & inventors;* p. 254–255.

Argon
CHEMISTRY

Normally a colorless, odorless gas, argon, atomic number 18, does not form stable compounds with other elements. Commercially-produced argon is obtained from the air by liquefaction and fractional distillation. Argon was discovered in 1894 by the scientists Lord Rayleigh (England) and Sir William Ramsay (Scotland). Chemical symbol is Ar.

Additional reading: *Van Nostrand's scientific encyclopedia;* p. 219.

Arithmetic
MATHEMATICS

Clay tablets have shown that the Babylonians were skilled in arithmetic as early as 4000 years ago. The Egyptians, around 1700 B.C., were also familiar with many concepts of arithmetic, according to an ancient document written at that time by the Egyptian priest Ahmes. They devoted much time to the handling of fractions and had symbols equivalent to our plus and minus signs. A Greek astronomer, Thales, studied with Egyptian priests around 600 B.C. One of his students may have been the Greek scholar Pythagoras. Thales is said to have introduced arithmetic to Europe.

Additional reading: Newman. *World of mathematics;* v. 1, p. 12, 79–82.

Arsenic
CHEMISTRY

Commonly existing as a steel-gray metal, arsenic, atomic number 33, can also be found in a reddish form (as arsenic sulfide). It was this latter form which the Greek scholar Aristotle observed, estimated to be about 350 B.C. It was not isolated until 1250 by Albert Magnus, a German scientist. It has many industrial uses. Chemical symbol is As.

Additional reading: *Van Nostrand's scientific encyclopedia;* p. 223–234.

Arsphenamine
PHARMACOLOGY

In 1909 Paul Ehrlich (Germany) discovered that the chemical compound arsphenamine was able to kill the spirochete microbe responsible for syphilis. Ehrlich named this drug Salvarsan 606 and announced its ability to cure syphilis in 1910. Ehrlich's breakthrough is considered to mark the starting point for the synthetic drug industry. He also coined the term "chemotherapy" at this time. In 1912 he introduced Neosalvarsan, which had the advantage of being water-soluble.

Additional reading: Ackerknecht. *A short history of medicine;* p. 232–233.

Artificial Ear Drum
MEDICINE

In 1640 Marcus Banzer (Germany) described the first use of an artificial ear drum for the tympanic membrane in his book entitled *De Auditione Laesa.* Banzer used a piece of swine's bladder connected to a tube fashioned from an elk's hoof. Joseph Toynbee, an English surgeon, devised an artificial ear drum in 1852 that consisted of a thin disk made from **vulcanized rubber** affixed to a rod.

Additional reading: Berger. *The hearing aid;* p. 38–40.

Artificial Heart
MEDICINE

In 1935 French-born surgeon Alexis Carrel (U.S.) and Charles Augustus Lindbergh (U.S.) designed an artificial heart used to keep blood passing through heart tissue. American heart surgeon Michael Ellis DeBakey designed a left ventricle in 1966 which he implanted in a patient. In 1969 American surgeon Denton Cooley implanted an entire heart made of Silastic with Dacron lining which had been designed by Argentine-born Domingo Liotta (U.S.). That heart was removed from the patient after 65 hours and replaced with a human heart. In 1982 Robert Koffler Jarvik, an American physician, designed a heart which was implanted in Barney Clark, an American dentist, who survived 112 days with his permanent artificial heart.

Additional reading: Robinson. *The miracle finders;* p. 189–212.

Artificial Heart Valve
MEDICINE

Charles A. Hufnagel (U.S.) developed the first artificial heart valve and used it on a human patient in 1952. The valve consisted of a small lucite tube with a float that rose and fell with the heart beat.

Additional reading: Robinson. *The miracle finders;* p. 169.

Artificial Hip
MEDICINE

The British orthopedic surgeon John Charnley designed the first satisfactory artificial hip in 1972. His successful innovation made use of a high-density polyethylene to structure the hip socket.

Additional reading: Robinson. *The miracle finders;* p. 96–99.

Artificial Insemination
PHYSIOLOGY

In 1780 the Italian biologist Lazzaro Spallanzani began performing artificial insemination experiments using amphibians, and in 1785 he successfully carried the process out on a dog. The next step forward came in 1901 when Russian biologist Ilya Ivanovich Ivanov founded the world's first center for the artificial insemination of horses. Ivanov's center was a success and by 1936 there were over 6 million cattle and sheep artificially inseminated in

Russia. The world's first "test tube" baby was born in England in 1978 as a result of in vitro fertilization and implantation of the embryo in the mother's uterus.

Additional reading: Magner. *A history of the life sciences;* p. 193–196.

Artificial Intelligence — COMPUTER SCIENCE

The concept of the possibility of artificial intelligence was introduced in a paper written by Alan M. Turing (England) in 1950. Artificial intelligence as a term was first used officially in 1956 at the initial conference on artificial intelligence. The field is concerned with devices which can "think" or function in a fashion similar to humans. In 1958 Frank Rosenblatt (U.S.) developed the Perceptron, an intelligent machine modeled after the human brain. Two related fields are **expert systems** and **robots.**

Additional reading: McCorduck. *Machines who think;* p. 57–61, 87–90.

Artificial Kidney — MEDICINE

John Jacob Abel, an American biochemist, produced the first useful artificial kidney for laboratory work in 1912. Dutch-born Willem J. Kolff (U.S.) began to work on building an artificial kidney that could be used on humans in 1938. Finally, in 1945, he designed a machine that worked to keep patients alive by filtering out blood urea, a process known as dialysis. Long-term dialysis became possible in 1960 when Belding Scribner (U.S.) developed a Teflon and Silastic shunt that could be left in a patient's wrist over a period of years. This shunt served to connect the artificial kidney machine to the patient.

Additional reading: Robinson. *The miracle finders;* p. 68–73.

Artificial Radioactivity — NUCLEAR PHYSICS

While studying the effect of bombarding aluminum with alpha particles in 1934, French physicists Frederic Joliot and Irene Joliot-Curie discovered that the aluminum emitted not only protons but also positrons. The unusual event, however, was the continued emitting of positrons even after the bombardment ceased. They realized they had created a new phosphorus, a version that did not occur in nature and was radioactive. Thus they had discovered artificial radioactivity. These new forms of radioactive elements, or isotopes, are now called **radioisotopes.** Thousands of these artificially created isotopes have been prepared for use in medicine and in industry.

Additional reading: Asimov. *Asimov's biographical encyclopedia;* p. 771–772.

Ascorbic Acid. *See* **Vitamin C.**

Aspirin. *See* **Acetylsalicylic Acid.**

Associative Psychology — PSYCHOLOGY

Associative psychology was founded by David Hartley (England) with the publication of his book entitled *Observations on Man.* Hartley suggested that ideas originated with sensory impressions and that sense organs received stimuli which in turn create vibrations in the nervous system. Hartley used principles of physiology to support his ideas.

Additional reading: Murray. *A history of Western psychology;* p. 91–93.

Astatine — CHEMISTRY

Occuring in nature only as a radioactive substance, astatine, atomic number 85, is found only in minute amounts as a result of the decay of uranium minerals. The total amount of astatine in the earth's crust is less than one ounce. It is the heaviest of the halogen group, and was first identified as an element in 1940 by the American scientists D. R. Corson, K. R. Mackenzie, and Emilio Segre. Chemical symbol is At.

Additional reading: *Van Nostrand's scientific encyclopedia;* p. 241.

Asteroid — ASTRONOMY

Around 1800 astronomers began a concerted effort to see if a planet existed between Mars and Jupiter. The German astronomer Heinrich Olbers led the search but it was an Italian astronomer, Giuseppe Piazzi, who inadvertently discovered what he thought was a small planet in that position in 1801. The search continued, and by 1807 three more small planets had been found. Sir William Herschel (England) suggested they be called asteroids because they appeared as starlike points of light rather than as disks, as planets do. By 1866 many asteroids had been discovered (by the 1980s over 1,600); then began a period when asteroids were found orbiting other planets, such as the discovery in 1920 by Walter Baade (Germany) of an asteroid that went as far from the sun as did Saturn.

Additional reading: Asimov. *Asimov's new guide to science;* p. 146–149.

Astrolabe — ASTRONOMY

An ancient astronomical instrument, the astrolabe, was used to predict the positions of the sun and stars. The earliest type was probably devised by the Greeks around 100 B.C., and was later developed by the Arabs. It can be set to show the appearance of the sky for any date or time. By the nineteenth century it had been replaced by mechanical clocks.

Additional reading: *Illustrated science and invention encyclopedia;* v. 2, p. 168–169.

Atabrine. *See* **Quinacrine.**

Atom Smasher. *See* **Particle Accelerator.**

Atomic Bomb. *See* **Nuclear Weapon.**

Atomic Clock — ELECTRONICS

The atomic clock is a timing device whose extreme accuracy is based on the undeviating frequency with which molecules vibrate. It was invented in 1953 by Charles H. Townes (U.S.) and is closely related to the **maser.** It makes use of the fact that the nitrogen atom in an ammonia molecule can be made to vibrate continuously at a constant rate of 24 billion times per second, doing so more accurately than any other timing device yet known. The vibrations can be made to regulate time measuring devices.

Additional reading: Asimov. *Asimov's new guide to science;* p. 451–452.

Atomic Theory — ATOMIC PHYSICS

The theory that all materials are composed of tiny, indivisible particles was proposed by John Dalton (England) in 1803 following his many experiments. He used the term "atom" (coined by a Greek philosopher named Democritus, around 450 B.C., whose conclusions were based on intuition, not experimentation). The theory of the atom as having a positive nucleus surrounded by electrons was developed by Ernest Rutherford (England) in 1911. This was refined by Niels Bohr (Denmark) in 1913 and by Arnold Sommerfeld (Germany) in 1916.

Additional reading: Asimov. *Asimov's biographical encyclopedia;* p. 11–12, 259–261, 624, 635–637, 701–702.

Audio Recorder. *See* **Magnetic Recording.**

Audiometer — ACOUSTICS

To aid his research on the mechanics of the cochlea, Hungarian-born physicist Georg von Bekesy (U.S.) developed an "audiometer" during the 1960s. Designed to test the hearing function, it was able to distinguish between deafness caused by a functional loss in the cochlea and that caused by a problem with the auditory nerve.

Additional reading: Sourkes. *Nobel prize winners in medicine and physiology 1901-1965;* p. 390–396.

Aureomycin. *See* **Antibiotic.**

Autoclave AGRICULTURE

The autoclave was invented in 1874 by A.K. Shriver (U.S.) and used steam to process cans needed for canning fruits and vegetables. The autoclave reduced both processing time and the danger of spoilage. (*See also* Can (Container))

Additional reading: Schlebecker. *Whereby we thrive;* p. 170–171.

Autogiro. *See* **Helicopter.**

Automatic Pilot AEROSPACE ENGINEERING

The automatic pilot is a device that automatically adjusts settings of an airplane's ailerons, rudder, elevator, etc., to keep the plane on a pre-set flying plan. The first one was invented in 1914 by Lawrence Sperry and his father Elmer Sperry (U.S.). A later addition was the autothrottle for automatic speed control.

Additional reading: *Illustrated science and invention encyclopedia;* v. 2, p. 206.

Automatic Transmission MECHANICAL ENGINEERING

In an automatic transmission system the driver does not have to select gears to use the one most suitable for the vehicle; it is done automatically. Most systems use a fluid for the medium to connect the engine to the gears turning the wheels. One of the earliest models was invented in 1910 in Germany. In 1926 Harold Sinclair (U.S.) applied one to an automobile, and in 1929 the German inventor Gottlieb Daimler developed the torque converter, which allows for a continuously variable gear ratio over a certain range.

Additional reading: *Illustrated science and invention encyclopedia;* v. 2, p. 200–205.

Automation CONTROL SYSTEMS

For centuries inventors devised various sorts of machines or gadgets that were self-acting, once started. For example, the ingenious water clock was built by the Greek inventor Ctesibius around 250 B.C. The Renaissance saw the development of spring-wound devices, and the Industrial Revolution was ushered in by serious inventions like the **steam pump** in 1698. The term "automation" was coined in 1947 by Delmar Harder (U.S.). The arrival of the computer brought rapid changes. One of the earliest computers built just for process control was made in 1957 by the Ramo-Woolridge Corporation (U.S.); such devices are now made by many firms around the world. (*See also* Robot)

Additional reading: Standh. *History of the machine;* p. 33–34, 202–203.

Automobile MECHANICAL ENGINEERING

The first automobile had three wheels, was powered by a steam engine, and had a top speed of about 3 miles per hour. It was invented in 1769 by the French artillery engineer Nicholas-Joseph Cugnot. Steam-powered automobiles were never popular due to their bulk and inefficiency. A vehicle powered by a mixture of air and hydrogen was tested by Samuel Brown (England) around 1820. The invention of the **internal combustion engine** encouraged inventors of cars. In 1860 Etienne Lenoir (France) built the first internal combustion engine and tested it in a carriage. Not until 1885 was a successful, three-wheel model built by the German engineer Karl Benz. Probably the first U.S. gasoline-driven car was built in 1893 by the Duryea brothers, Charles and Frank.

Additional reading: Feldman. *Scientists & inventors;* p. 164–165, 180–181, 200–201.

Automobile Brake. *See* **Brake.**

Automobile Carburetor MECHANICAL ENGINEERING

A device for mixing air and gasoline for internal combustion machines is called a carburetor. In 1888 Edward Butler (England) patented a spray carburetor, and in 1893 Wilhelm Maybach (Germany) invented a carburetor that was used on a two-cylinder automobile engine. That same year Karl Benz (Germany) invented a butterfly valve that regulated the fuel-air mixture.

Additional reading: *Illustrated science and invention encyclopedia;* v. 4, p. 478–479.

Automobile Engine. *See* **Internal Combustion Engine.**

Automobile Ignition System ELECTRICAL ENGINEERING

The ignition system is needed to provide sparks to ignite fuel-air mixtures in cylinders of internal combustion engines. It depends upon several components. The induction coil was invented in 1841 by the French physicists Antoine Masson and Louis Brequet, being perfected in 1851 by the German physicist Heinrich Ruhmkorff. The combined use of a storage battery and induction coil was devised in 1883 by Etienne Lenoir (France) as well as by Karl Benz (Germany). Lenoir also invented the spark plug in 1885. The magneto, which converts mechanical energy into electrical, was invented in 1890 by Robert Bosch (Germany) and used on an automobile in 1897. In recent years an electronic ignition system has been invented.

Additional reading: *Illustrated science and invention encyclopedia;* v. 9, p. 1243–1245.

Automobile Self-Starter MECHANICAL ENGINEERING

Early automobilies had to be cranked by hand, a cumbersome method that could be hazardous. In 1912 the American inventor Charles F. Kettering invented a successful self-starter powered by the car's storage battery.

Additional reading: Asimov. *Asimov's new guide to science;* p. 437.

Automobile Speedometer MECHANICAL ENGINEERING

The principles behind the speedometer were first studied by the English mathematician Charles Babbage around 1840. It was not until 1904 that a governor-type speedometer was devised by L. E. Cowley (England). In 1911 W. H. Grossmann (Germany) invented a mechanical speedometer with an accompanying speed-indicator. Most speedometers measure the speed of rotation of the automobile transmission, with a flexible cable attached to a pointer on an indicator display. Closely related is the distance-measuring instrument called the odometer, which is driven by gears attached to the speedometer spindle.

Additional reading: *Illustrated science and invention encyclopedia;* v. 16, p. 2190–2191.

Automobile Tire. *See* **Rubber Tire.**

Automobile Transmission MECHANICAL ENGINEERING

The devise for transferring power from the engine to the drive shaft is called the transmission. A shaft runs from the motor to the drive wheels, where the differential gear transfers power to the wheels. One of the earliest models was devised by the French inventer Louis Renault in 1899, offering three forward speeds and one reverse. (*See also* Automatic Transmission)

Additional reading: *Illustrated science and invention encyclopedia;* v. 4, p. 462–464.

Avogadro's Number CHEMISTRY

The Italian physicist Amedo Avogadro made serious studies of atoms and molecules (a word he coined), particularly of their weight. He contended that there is a fixed number of atoms or molecules in any gas in which the number of grams of the gas is numerically equal to its atomic or molecular weight. Although he proposed this around 1811, it was not generally accepted until 1858 when the Italian chemist Stanislao Cannizzaro showed its validity.

Additional reading: Asimov. *Asimov's biographical encyclopedia;* p. 277–278, 439–440.

Azathioprine. *See* **Kidney Transplant.**

B

B-Vitamin Complex. *See* **Vitamin B.**

Bacteria MICROBIOLOGY

Anton van Leeuwenhoek, a Dutch biologist and microscopist, discovered bacteria in the human mouth in 1683 while using a single lens **microscope** of his own design. In 1773 Danish biologist Otto Friedrich Muller was able to see bacteria well enough to assign two categories, namely bacillum and spirillum. The science of bacteriology was founded by German botanist Ferdinand Julius Cohn in 1872 with the publication of his 3-volume treatise on bacteria. Cohn is also given credit for coining the term "bacterium" from the Latin word meaning "little rod." During the 1880's Robert Koch (Germany) founded modern medical bacteriology, while in 1884 Danish bacteriologist Hans Christian Joachim Gram established a way to distinguish Gram-positive bacteria.

Additional reading: Magner. *A history of the life sciences;* p. 237–287.

Bacteriophage VIROLOGY

English bacteriologist Frederick Twort discovered that a virus had infested and killed his bacteria in 1915. Canadian-born Felix Hubert d'Herelle independently made the same discovery in 1917 while working at the Pasteur Institute in Paris. He named the virus "bacteriophage." In 1942 Salvador Edward Luria (U.S.) obtained the first high quality electron micrograph of a bacteriophage. Alfred Day Hershey, an American microbiologist, showed in 1952 that the **nucleic acid** of the phage entered the cell, carrying the genetic material. Luria, Hershey, and Max Delbruck (U.S.) shared the 1969 Nobel Prize in medicine and physiology for their research with bacteriophages and bacterial genetics.

Additional reading: Watson. *The double helix;* p. 1–131.

Bailey Bridge CIVIL ENGINEERING

Military bridges were found to be too light in World War II to hold heavy armored vehicles. In 1941 the British inventor Sir Donald Bailey devised a bridge made up of as many sections as needed. Piers can be erected at intervals to support sections spanning wide gaps. For crossing rivers the sections floated on pontoons.

Additional reading: *Illustrated science and invention encyclopedia;* v. 2, p. 220.

Bakelite MATERIALS

About 50 years after the introduction of **celluloid,** the Belgian-born chemist Leo Baekeland discovered in 1909 a new substance while experimenting with formaldehyde. Upon heating it became soft, after which it could be molded and then hardened; it could also be made in powdered form, set under pressure, and heated to form a hard solid material. Nonconductive and resistant to heat and corrosion, it was named Bakelite, the first synthetic plastic. Many products were made from it, so it can be said to have triggered the start of the plastics industry.

Additional reading: Feldman. *Scientists & inventors;* p. 246–247.

Balance (Chemical) CHEMISTRY

The earliest balances consisted of a beam pivoted at its center with pans at either end for holding standard weights and the material to be weighed. Even the simplest ones, probably used by the Egyptians around 5000 B.C., were quite accurate. The Romans invented the "knife edge" fulcrum on which the beam balanced, increasing its accuracy. In laboratory use the balance is usually enclosed in a glass box, thus eliminating dirt and air currents. Both single and double pan models exist. Sensitive balances are called microbalances and can weigh to a millionth of a gram. Electrical and electronic circuits are involved.

Additional reading: *Illustrated science and invention encyclopedia;* v. 2, p. 222–224.

Ballistic Missile. *See* **Missile.**

Balloon AEROSPACE ENGINEERING

A desire for manned flight first came to fruition in 1783 when French papermakers Jacques and Joseph de Montgolfier rose nearly a mile in a balloon made of linen, heated by a fire of coal and straw. By 1804 a manned balloon carrying the French scientist Joseph Gay-Lussac reached a height of more than four miles. In later years efforts to direct the flight of balloons led to the development of **airships.** Interest in balloons increased in 1931 when Auguste Piccard (France) and his brother Jean rose to a height of 11 miles inside a cabin, attached to a balloon. By the 1980s manned balloons had soared to more than 23 miles.

Additional reading: Asimov. *Asimov's new guide to science;* p. 211–212.

Ballpoint Pen. *See* **Pen.**

Barbed Wire AGRICULTURE

One of the important means for developing the American plains areas was the simple barbed wire fence, making possible the separation of field crops from cattle. Barbed wire was patented in the United States in 1867. The first practical machine for making such wire was patented in 1874 by the American farmer Joseph Glidden. Usually high tensile steel is used.

Additional reading: *Illustrated science and invention encyclopedia;* v. 2, p. 236–237.

Barbiturate PHARMACOLOGY

In 1863 German chemist Johann Friedrich Wilhelm Adolf von Baeyer discovered barbituric acid, which was the parent compound of barbiturates, a group of ureides that act as central nervous system depressants. In 1911 German chemist Emil Hermann Fischer synthesized phenobarbital, a sleeping pill that turned out to be an effective anticonvulsive drug. H. A. Shonle (U.S.) and A. Moment (U.S.) synthesized amylobarbitone in 1923. The Eli Lilly Company marketed it as Amytal. In 1929 Shonle also synthesized quinalbarbitone which Lilly marketed as Seconal. (*See also* Sodium Barbital)

Additional reading: Sneader. *Drug discovery;* p. 24–34.

Barium CHEMISTRY

A soft, ductile metal, barium, atomic number 56, occurs normally as a sulfate. It was discovered in 1808 by the English scientist Sir Humphry Davy. Georgia and Tennessee are the main states producing it. Chief use to date has been that of eliminating oxygen in electronic vacuum tubes. Chemical symbol is Ba.

Additional reading: *Van Nostrand's scientific encyclopedia;* p. 303–304.

Barograph. *See* **Barometer.**

Barometer METEOROLOGY

Curiosity about the nature of vacuums led the Italian physicist Evangelista Torricelli to experiment with columns of mercury. From that work came the mercury barometer in 1644, which allowed for measurement of the varying weight of the atmosphere as the weather changed at a given place. A device for recording atmospheric pressure (barograph) was not invented until 1681, primarily by the English physicist Robert Hooke, who improved on a device developed in 1683 by the English architect Sir Christopher Wren.

Additional reading: Feldman. *Scientists & inventors;* p. 38–39.

Base (Chemistry)
CHEMISTRY

A base is a chemical compound capable of accepting a proton (hydrogen ion) from another substance. The greater the tendency to accept a proton, the stronger the base. Although chemists studied bases for centuries, one of the first to propose a measure of their strength was the Swedish chemist Svante Arrhenius in a paper published in 1884 that was chiefly concerned with electrolytes. In 1923 the Danish scientist Johannes Bronsted proposed a definition of **bases** and **acids.** That same year the Danish chemist Niels Bjerrum devised a method of using the strength constant of acids and bases to study the dissociation of certain chemicals.

Additional reading: *Dictionary of scientific biography;* v. 2, p. 170, 499; v. 6, p. 298–299.

Basic Oxygen Process
METALLURGY

An improvement on the open-hearth process of steelmaking is called the basic oxygen process. It was developed around 1950 in Lin-Donawitz, Austria, and has subsequently been known as the L-D process as well as the basic oxygen process. The equipment resembles a Bessemer converter except that an oxygen blast comes from above, not beneath. It offers greater speed and lower installation costs than the open-hearth process. It can produce all types of steel and has become the major method used in the United States.

Additional reading: *Illustrated science and invention encyclopedia;* v. 17, p. 2240–2241.

Bathyscaphe
OCEANOGRAPHY

Need for a way to explore the ocean at great depth led the Swiss physicist Auguste Piccard to invent the bathyscaphe in 1948. It consists of a hull-shaped float and a spherical steel cabin suspended beneath it. Dives as deep as 35,000 feet have been made in the Pacific Ocean.

Additional reading: *Illustrated science and invention encyclopedia;* v. 2, p. 245–249.

Bathysphere
OCEANOGRAPHY

A desire to study coral at great depths prompted the American naturalist William Beebe to build around 1934 a strong diving device capable of deep dives. He did so with the help of a colleague, Otis Barton (U.S.), diving to a depth of 3,000 feet. A better device was the **bathyscaphe.**

Additional reading: Asimov. *Asimov's biographical encyclopedia;* p. 662–663.

Battery. *See* Electric Battery.

Beaufort Wind Scale
OCEANOGRAPHY

For recording the strength of winds as observed by ships at sea, a scale was needed to make observations as consistent as possible. In 1805 the hydrographer Admiral Sir Francis Beaufort (England) devised a scale which bears his name. It ranges from calm to hurricane force. These ranges were used on weather maps for 150 years, being replaced in 1955 by symbols based on wind speed in knots. His chart included a list of effects on both land and sea to help estimations of wind speeds. His scale is closely related to symbols for the state of the sea. (*See also* Douglas Sea Scale; Petersen Scale)

Additional reading: Fairbridge. *Encyclopedia of oceanography;* p. 786–792, 991–992.

Behaviorism
PSYCHOLOGY

J. B. Watson (Germany) began a school of psychology known as behaviorism in 1914 when he published the first paper on it. Watson advocated the study of observable behavior.

Additional reading: Murray. *A history of Western psychology;* p. 258–286.

Benzene
ORGANIC CHEMISTRY

The creation of a plan for showing the **chemical structure** of compounds by the German chemist Friedrich Kekule prompted him to study carbon and its valency, which he published in 1858. From this he turned to investigate the structure of benzene, an important hydrocarbon. The unknown structure puzzled chemists and had prevented the development of synthetic dyes. In 1865, during what Kekule later described as "a half sleep," he realized the molecule had to have a ring of six carbon atoms, each attached to a hydrogen atom. This proved to be correct, and it was the starting point for our understanding of all the many organic structures.

Additional reading: Feldman. *Scientists & inventors;* p. 176–177.

Benzodiazepine. *See* Librium.

Berkelium
CHEMISTRY

A radioactive metal of the actinide series, berkelium, atomic number 97, is also one of the transuranium elements. It is produced by bombardment of certain heavy elements, such as americum. It was discovered in 1949 by the American scientists Glenn T. Seaborg, A. Ghiorso, and S. G. Thompson. Chemical symbol is Bk.

Additional reading: *Van Nostrand's scientific encyclopedia;* p. 336–337.

Beryllium
CHEMISTRY

Metallic beryllium, atomic number 4, was not produced until 1828, three decades after its discovery in 1798 by the French scientist L. N. Vauquelin as a constituent of the minerals beryl and emerald. A mill in Utah is the only one which commercially extracts beryllium from its ores. Beryllium has many uses in the aerospace industry because of its low density and high thermal capacity. Chemical symbol is Be.

Additional reading: *Van Nostrand's scientific encyclopedia;* p. 339–340.

Bessemer Converter
METALLURGY

Steelmaking was aided by the introduction of the Bessemer converter, an invention of the English engineer Sir Henry Bessemer. In 1856 he began commercial use of his converter, which was a pear-shaped container holding molten pig iron into which a blast of air was forced. The effect was to form a slag (consisting of the impurities in the iron) which could be easily renewed. Afterwards carbon and other ingredients could be added to produce steel. This process made mass production possible, reducing the cost of steel, and increasing its use. Six years earlier the American ironmaster William Kelly had devised the same basic process, but he did not patent it until 1857. In later years other methods included the **open-hearth process** and the **basic oxygen process.**

Additional reading: *Illustrated science and invention encyclopedia,* v. 3, p. 278–280.

Beta Particle
NUCLEAR PHYSICS

The process of **radioactivity** was shown in 1900 by Antoine Becquerel (France) to consist partly of particles identical to the **electron.** In 1900 the British physicist Ernest Rutherford named them beta rays, although they are usually now spoken of as beta particles. He also did work on other parts of the emission; namely, **alpha particles** and **gamma rays.**

Additional reading: Asimov. *Asimov's new guide to science;* p. 317–318.

Betatron
NUCLEONICS

Problems with accelerating electrons included their increase of mass as they neared the speed of light. In 1940 a new accelerator for electrons was designed by the American Donald Kerst. Named the betatron, after **beta particles**, it automatically balanced increasing mass with an increasing electrical field to keep their speed high.

Additional reading: Asimov. *Asimov's new guide to science;* p. 339.

Bevatron. *See* **Synchrotron.**

Big Bang Theory. *See* **Universe's Origin.**

Binary Star
ASTRONOMY

The discovery by the English astronomer John Goodricke in 1782 that the star Algol periodically got bright, then dim, caused him to theorize that a companion star regularly passed in front of Algol, thus eclipsing it and causing its light to dim. This was not confirmed until 1889 when the German astronomer Hermann Vogel determined spectroscopically that Algol's spectral lines showed color shifts associated with receding and approaching motion. Thus binary stars were accepted as a new type. In 1890 Vogel discovered spectroscopic binaries, two stars so close together they appear as a single star even in good telescopes.

Additional reading: Asimov. *Asimov's new guide to science;* p. 41–42.

Biology
BIOLOGY

Karl Friedrich Burdach (Germany) introduced the word biology in 1800 to indicate the study of the morphology, physiology, and psychology of human beings. In 1802 Gottfried Treviranus, a German naturalist, and Jean Baptiste Lamarck, a French zoologist, in separate publications, broadened the meaning of biology to include the study of all life.

Additional reading: Magner. *A history of the life sciences;* p. 1–5, 341–342.

Birth Control Pill
MEDICINE

Russel E. Marker, an American chemist, discovered a method for synthesizing large quantities of the female sex hormone progesterone in 1939. In 1951 another American chemist, Carl Djerassi, following Marker's methodology, synthesized a new and powerful equivalent to progesterone called 19-Norsteroids. In the same year, American physician John Rock, American biologist Gregory Goodwin Pincus, and Chinese-born biologist Min Chuch Chang (U.S.) found that 19-Norsteroids suppressed ovulation. They had thus developed the first birth control pill.

Additional reading: Robinson. *The miracle finders;* p. 265–273.

Bismuth
CHEMISTRY

A silvery-white metal, bismuth, atomic number 83, is a naturally occurring substance in lead and copper ores. It was first described by the German chemist Basil Valentine around 1604, but it was not defined as a new element until 1753, when its properties were published by the French chemist Claude Geoffroy, the German chemist Johann Pott, and the Swedish chemist Torbern Bergman. It is used chiefly in cosmetics, pharmaceuticals, and industrial chemicals. Chemical symbol is Bi.

Additional reading: *Van Nostrand's scientific encyclopedia;* p. 374–375.

Black Body Radiation
PHYSICS

The study of how objects emit radiation, such as heat, led to the law put forth by the Austrian physicist Josef Stefan in 1879 that the total radiation emitted in a perfectly black body depended upon its temperature, not its physical makeup. The wave length of the radiation was found by the German physicist Wilhelm Wien in 1893 to fit Stefan's formula. Then in 1900 the German physicist Max Planck showed that all radiation is emitted in small units, leading to his **quantum theory.**

Additional reading: Asimov. *Asimov's new guide to science;* p. 386–387.

Black Hole
ASTRONOMY

Black holes are believed to be aging stars whose structure has collapsed so that their volume approaches zero while their gravitational force on their surface approaches infinity; thus their density is extremely high (1.5 million tons per cc). Only about one visible star in a thousand is large enough to have such a collapse, but we have no way of knowing how many others have already collapsed. A typical sequence of deterioration of stars is to collapse and become a **white dwarf star,** then to become a **neutron star,** then to become a black hole. The term was first used by the American physicist John A. Wheeler in the 1960s. What is believed to be the first black hole was discovered by the American astronomer C. T. Boltin in 1971.

Additional reading: Buedeler. *The fascinating universe;* p. 227–229.

Blacktop Road
CIVIL ENGINEERING

The use of crushed rock on **macadam roads** in the early nineteenth century was reduced with the introduction of rubber tires on automobiles. It was necessary to mix tar or bitumen with the stones to preserve the surface and to produce a smoother surface. Although the person responsible for introducing the use of the tar or asphalt binder is not known, the basic concept of crushed stone goes to the Scottish surveyer John L. McAdam and his work in 1823. The present system consists of heating the mix of stone and bitumen and delivering the hot mix to the site, where it is compacted by road rollers.

Additional reading: *Illustrated science and invention encyclopedia;* v. 15, p. 1989–1993.

Blast Furnace
METALLURGY

The use of a blast of air to facilitate the melting of iron ore was known as far back as 2200 B.C. At first the blasts were provided by hand-operated bellows. Later short chimneys were built so that a better blast could be achieved. Furnaces of this sort have been found in Sumerian and Chinese ruins, dating back to 1400 B.C. Fuels changed from wood to coke in 1709 due to work done by Abraham Darby (England). The main innovation involving the blast furnace came in 1856 when Henry Bessemer (England) discovered that the furnaces got much hotter by sending the blast of air through holes in the bottom of the heating vessel. (*See also* Bessemer Converter)

Additional reading: *Illustrated science and invention encyclopedia;* v. 3, p. 290–292.

Bleaching
TEXTILES

Bleaching textiles in early eras was probably simply a process of placing them in the sun. The use of buttermilk lasted for centuries until 1758, when dilute sulphuric acid was used, still coupled with exposing items to the sun. In 1785 chemical bleaching became possible due to the discovery by the French chemist Claude Berthollet that a solution made by passing chlorine through potash was suitable as a bleach. In 1799 the English chemist Smithson Tennant discovered that passing chlorine over lime was successful, becoming a great aid to the cotton and paper industries. An improved method was devised in 1870 by Henry Deacon (England) using a mixture of chlorine and hydrochloric gases. In 1886 the electrochemical production of sodium peroxide was invented by Hamilton Castner (U.S.).

Additional reading: Derry. *Short history of technology;* p. 265, 536–537, 541.

Blood Circulation
PHYSIOLOGY

In 1553 Miguel Serveto (Spain) published *Christianismi Restitutio* in which he asserted that blood passes from the heart to the lungs and back to the left ventricle of the heart. Around the same time, in 1559, Realdo Colombo (Italy) reached the conclusion that blood passes from the right side of the heart into the lungs and then to the left ventricle, thus describing what is known as the "smaller circulation." In 1628 William Harvey, the English physician, published *Exercitatio De Motu Cordis et Sanguinis* in which he described the flow of blood, including the idea that the heart works as a pump to send blood to the arteries.

Additional reading: Butterfield. *The origins of modern science;* p. 49–66.

Blood Groups IMMUNOLOGY

Human blood groups were discovered in 1900 when Austrian-born physician Karl Landsteiner (U.S.) demonstrated the existence of AB0 blood groups. In 1902 Landsteiner discovered the AB group and in 1927 the M and N groups. Landsteiner found that the groups differed according to antigens, proteins which occurred in the blood. He also showed that the A group had an anti-B antibody and vice versa; AB had no antibodies; and O had no antigens. By establishing this pattern, Landsteiner made it possible to proceed safely and effectively with procedures for **blood transfusion.** His work also enabled the first blood bank to be founded. Landsteiner was awarded the 1930 Nobel Prize for medicine and physiology for the discovery of human blood groups.

Additional reading: "20 discoveries that changed our lives"; p. 65–67.

Blood Pressure. *See* **Kymograph.**

Blood Transfusion MEDICINE

Richard Lower, an English physician, was the first to perform a direct blood transfusion from one animal to another in 1665. Soon thereafter, in 1667, French physician Jean Baptiste Denis transfused the blood of a lamb into an ill young man who survived long enough for Denis to perform another such transfusion into a different patient. When both men died, Denis was charged and then acquitted of murder. Transfusions of this type were ordered forbidden. In 1900 Austrian-born Karl Landsteiner (U.S.) made transfusions safe when he discovered human **blood groups.**

Additional reading: Feldman. *Scientists & inventors;* p. 256–257.

Blood Vessel ANATOMY

Alcmaeon, a Greek physician born around 535 observed that veins were blood vessels and were different from arteries. He did not realize that arteries also functioned as blood vessels since those he studied were from cadavers and hence were empty of blood. Praxagoras, a Greek physician born around 340 B.C., was the first to realize that both veins and arteries carried blood. He is given credit for discovering that two types of blood vessels existed.

Additional reading: Asimov. *Asimov's new guide to science;* p. 597–599.

Blower. *See* **Supercharger.**

Boat NAVAL ARCHITECTURE

Simple boats made out of hollowed-out logs (dugout canoes) were in use in Africa and other regions around 6000 B.C. Flat bamboo rafts were in use in China starting around 4000 B.C. Boats made of frames covered with skins or bark existed in Mesopotamia as far back as 3000 B.C. Around 1000 B.C. the Chinese began to replace the bamboo with planks, making the sampan. (*See also* Oar; Ship)

Additional reading: *Illustrated science and invention encyclopedia;* v. 3, p. 297–300; v. 20, Section E.

Boiler. *See* **Steam Heating; Steam Pump.**

Boolean Algebra MATHEMATICS

In 1854 the English mathematician George Boole wrote a book on symbolic logic which noted that the statements used in deductive logic could be handled by using mathematical expressions. Thus his work aided the design of computers by forming the basis for computer operations such as information retrieval. His contribution is now known as Boolean algebra.

Additional reading: Asimov. *Asimov's new guide to science;* p. 859–860.

Boron CHEMISTRY

Boron, atomic number 5, is found either as a yellowish-brown crystalline solid or as an amorphous greenish-brown powder. It is used in the manufacture of certain semiconductor products and also as a means of increasing the strength of metals and alloys. It has been known since about 4000 B.C., starting with the Babylonians, but it was not until 1807 that the English scientist Sir Humphry Davy first produced elemental boron (in amorphous form). Chemical symbol is B.

Additional reading: *Van Nostrand's scientific encyclopedia;* p. 410–412.

Bose-Einstein Statistics. *See* **Boson.**

Boson NUCLEAR PHYSICS

Most subatomic particles can have one of two kinds of **particle spin.** Those with a spin that can be expressed in whole numbers are called bosons. They can be handled by rules worked out in 1924 by the Indian physicist Satyendranath Bose and the American physicist Albert Einstein, known as Bose-Einstein statistics. Those with the other type of spin are called **fermions.**

Additional reading: Asimov. *Asimov's new guide to science;* p. 342.

Botany BOTANY

Credit for founding botany, the science that studies plants and plant life, is given to Theophrastus, a Greek who lived from around 372 B.C. to 287 B.C. Theophrastus was in charge of the Lyceum in Athens after Aristotle retired. He carried on the traditional studies of biology and concentrated mainly on the plant world. He described over 550 species. In the 1500s, a German botanist, Leonhard Fuchs, wrote the first modern glossary of botanic terms.

Additional reading: Sarton. *A history of science;* p. 547–558.

Bottle Machine. *See* **Glass.**

Brake MECHANICAL ENGINEERING

Brakes have been in use since the invention of moving machinery and various means of transportation. They operate on air, steam, metallic friction, or hydraulic force. One of the earliest brakes was a steam brake, invented by Robert Stephenson (England) in 1833. In 1844 James Nasmyth (Scotland) invented the air brake. An automatic air brake for railroad cars was devised in 1869 by George Westinghouse (U.S.), and tried out for the first time on a passenger train in 1872. Brake shoes for automobiles were invented by Louis Renault (France) in 1902. In 1904 hydraulic brakes were invented by F. H. Heath (England) for use on cars.

Additional reading: *Illustrated science and invention encyclopedia;* v. 3, p. 325–329.

Brass METALLURGY

An alloy made from copper and zinc, brass was probably made as long ago as 1000 B.C. by people in northeastern Turkey. They made brass by heating copper with charcoal and powdered zinc ore. The Persians are known to have used it around 500 B.C. Brasses are classified according to the amount of zinc they contain. Different properties, such as tensile strength, corrosion resistance, or decorative appearance are obtained by controlling the type and quality of constituents added.

Additional reading: *Illustrated science and invention encyclopedia;* v. 3, p. 334–338.

Brick MATERIALS

Bricks have been used as a building material for thousands of years. The earliest example is that of clay bricks found on sites in Mesopotamia built around 2500 B.C. The Romans widely used bricks in Europe; however, after the fall of the Roman Empire there was a cessation for several centuries. Bricks were handmade until 1619, when a patent was granted for a clay working machine, but it was not until 1858 that a German engineer named Hoffmann invented the continuous kiln so as to permit speedier production. Today complete automation is common in modern manufacturing. The type of clay used depends upon the uses and properties expected.

Additional reading: *Illustrated science and invention encyclopedia;* v. 3, p. 349–352.

Bridge
CIVIL ENGINEERING

The earliest bridges were built of timber, leaving no trace for later centuries. Nabopolassar (Babylonia) built a stone bridge resting on brick piers across the Euphrates around 600 B.C. Trajan's bridge over the Danube, built around 99 A.D., had one semicircular arch with a span of 65 feet. In Italy in the fourteenth century a duke of Milan built a bridge over the River Adda having a 236-foot arch, a feat not equalled for more than 400 years. The English engineer John Rennie built three famous masonry bridges—in 1817 he finished London Bridge, having one span of more than 150 feet. The first all-iron bridge was built in 1779 by Abraham Darby (England). The basic types are beam, arch, and **suspension bridges**.

Additional reading: *Illustrated science and invention encyclopedia;* v. 3, p. 353–361.

Bromine
CHEMISTRY

Bromine, atomic number 35, is one of two elements that are liquid at normal temperatures and pressures (the other one is **mercury**). It is a dense red-brown liquid which vaporizes readily, having a very irritating odor. In the U.S. most of it is derived from natural brines in Arkansas and Michigan. It was discovered in 1826 by the French chemist Antoine-Jerome Balard. Its chief use is as an anti-knock fluid for motor fuels. Chemical symbol is Br.

Additional reading: *Van Nostrand's scientific encyclopedia;* p. 455–456.

Bronze
METALLURGY

The first definition of bronze was that of a copper alloy containing about 25% tin, but currently bronze can include alloys with little or no tin. Bronze has been used since about 3000 B.C., the beginning of the Bronze Age in some areas. Ancient Turkey was one site of early usage, perhaps due to a supply of tin found in that area. Molds for making bronze ornaments and domestic utensils were made of baked clay. Different constituents for controlling properties typically include phosphorus, lead, aluminum and zinc.

Additional reading: *Illustrated science and invention encyclopedia;* v. 3, p. 368–370.

Bronze Age. *See* **Copper Mining.**

Brownian Motion
ATOMIC PHYSICS

The random movement of tiny particles suspended in a fluid is called Brownian motion, named after the discoverer, the Scottish biologist Robert Brown, who observed it in 1827. In 1905 Albert Einstein (U.S.) developed a mathematical analysis of Brownian motion. He stated that the water molecules are continually bombarding all objects in the water. The smaller the object, the greater the number of molecules striking it. This phenomenon showed the effect of individual molecules. In 1913 the subject was further studied by the French physicist Jean Perrin, who used Einstein's theory to calculate the approximate size of the water molecules.

Additional reading: Asimov. *Asimov's biographical encyclopedia;* p. 271–272, 631–632, 674.

Bubble Memory
COMPUTER SCIENCE

A different type of memory device for computers was developed in 1967 at Bell Laboratories by a team led by A. H. Bobeck (U.S.). It consists of certain crystals which, when placed in a magnetic field, store magnetic signals (0 or 1). They are nonvolatile and can pack about 11 times as much data per unit volume as semiconductors.

Additional reading: *Illustrated science and invention encyclopedia;* v. 22, p. 21.

Bulk Modulus. *See* **Modulus of Elasticity.**

Bunsen Burner
CHEMISTRY

A basic piece of laboratory equipment is the Bunsen burner, which produces a flame with very little light yet adequate for heating chemicals. It was developed by the German chemist Robert Bunsen in 1857 in conjunction with his German technician C. Desaga. Various improvements have been made subsequently.

Additional reading: *Illustrated science and invention encyclopedia;* v. 3, p. 397–398.

Buoyancy
PHYSICS

The discovery of the principles of buoyancy, governing the force by which bodies are supported in liquids, is credited to Archimedes, a Greek mathematician and engineer. His success in using buoyancy as a means of determining the purity of gold in a king's crown is but one of the many stories about his accomplishments. Exact dates of his discoveries are not exactly known, but around 250 B.C. is a reasonable estimate.

Additional reading: Downs. *Landmarks in science;* chap. 5.

Butterfat Test
AGRICULTURE

Stephen M. Babcock (U.S.) invented the butterfat test in 1890 and marketed it in 1891. By using acid, heat, and centrifugal force, Babcock was able to precipitate a measurable percentage of butterfat. The advent of the test allowed butter factories to begin paying dairy farmers according to the amount of butterfat in their product for the first time.

Additional reading: Schlebecker. *Whereby we thrive;* p. 183–184.

Buys Ballot Law
METEOROLOGY

Around 1790 the German astronomer and physicist Heinrich W. Brandes noticed that when standing with one's back to the wind, higher air pressure was to the right and lower pressure to the left (in the Northern Hemisphere). In 1857 C. H. D. Buys Ballot (Holland) formulated this as a rule and made it widely known among sailors, for whom it was a great help for many years before more precise methods of weather analysis became available.

Additional reading: Fairbridge. *Encyclopedia of oceanography;* p. 990.

C

Cable. *See* **Submarine Cable; Wire Rope.**

CAD. *See* **Computer-Aided Processes.**

Cadmium
CHEMISTRY

A soft silver-white metal, cadmium, atomic number 48, is a relatively rare element. It was discovered in 1817 by the German metallurgist F. Stromeyer. It is found chiefly in zinc deposits. Its major use is for plating iron and steel to provide a protective coating, but it is also used to increase the hardness of alloys. Chemical symbol is Cd.

Additional reading: *Van Nostrand's scientific encyclopedia;* p. 482–483.

CAE. *See* **Computer-Aided Processes.**

CAI. *See* **Computer-Aided Processes.**

Calciferol. *See* **Vitamin D.**

Calcium
CHEMISTRY

Calcium, atomic number 20, is a silver-white metal and one of the most widely distributed elements. It was discovered by the English chemist Sir Humphry Davy in 1808. It is stable in dry air but in the presence of moisture emits hydrogen gas. The quantity mined and consumed is said to be the largest of any industrial chemical. It is widely used to improve the physical properties of iron and steel, as well as a component of many alloys. Chemical symbol is Ca.

Additional reading: *Van Nostrand's scientific encyclopedia;* p. 484–486.

Calculating Machine
COMPUTER SCIENCE

Use of mechanical devices to make calculations dates back to the **abacus.** A mechanical calculator was built by the German mathematician Wilhelm Schickard in 1623 using a set of metal wheels. Around 1642 the French mathematician Blaise Pascal made a similar machine that could handle up to nine-digit numbers. In 1673 the German mathematician Gottfried Leibniz made a device similar to Pascal's. A steam-powered digital calculator was proposed in 1823 by the English inventor Charles Babbage, but was never built, though his concepts were later useful. Around 1880 Herman Hollerith (U.S.) built a card-driven electromechanical tabulator. Beginning in the 1960s electronic calculators and **digital computers** became very popular.

Additional reading: Ritchie. *The computer pioneers;* p. 17–21.

Calculus
MATHEMATICS

Calculus is a branch of mathematics that allows for the solution of certain complex problems; it has two main forms—integral calculus and differential calculus. It was invented by the English scientist Sir Isaac Newton around 1670 and at about the same time by the German mathematician Gottfried W. Leibniz who wrote about the subject in 1684. His terminology and notation system were generally considered superior to that of Newton.

Additional reading: Asimov. *Asimov's biographical encyclopedia;* p. 150, 156.

Calendar
ASTRONOMY

The Egyptian calendar was introduced around 4242 B.C. and aided agriculture by enabling Egyptians to predict the date of the annual flood of the Nile. The length of their year was 365 days, with no consideration made for leap years.

Additional reading: Curwen. *Plough and pasture;* p. 28–29.

Californium
CHEMISTRY

A radioactive metal of the actinide series, californium, atomic number 98, is prepared only by nuclear synthesis. It was discovered in 1950 by S. G. Thompson, K. Street, Jr., A. Ghiorso, and Glenn T. Seaborg (all U.S.). It is a convenient and portable source of neutrons for scientific work and in various types of oil well logging and prospecting. Chemical symbol is Cf.

Additional reading: *Van Nostrand's scientific encyclopedia;* p. 491.

Caloric. *See* **Heat.**

Calorimeter
PHYSIOLOGY

Two Germans, Karl von Voit and Max Joseph von Pettenkofer, devised the first calorimeter large enough to hold human beings around 1860. Their calorimeter made it possible for the first time to measure oxygen consumed, carbon dioxide liberated, and heat produced. The human basal metabolic rate could be measured.

Additional reading: Asimov. *A short history of biology;* p. 88–90.

CAM. *See* **Computer-Aided Processes.**

Camera. *See* **Instant Camera; Photography; Video Camera.**

Can (Container)
MECHANICAL ENGINEERING

The manufacture of cans dates back to 1795 when the French confectioner Francois Appert found that food could be preserved by heating. In 1810 the French inventor Pierre Durand was granted a patent for using tin-plated steel for containers. Soon afterward a cannery was set up in London. In 1849 Henry Evans (U.S.) invented a machine which stamped out tops and bottoms. By the end of the nineteenth century cans were made by machines, not by hand. Both tinplate and aluminum have been used for cans of various sizes and shapes. Cans are lacquered internally to prevent contamination of the contents.

Additional reading: *Illustrated science and invention encyclopedia;* v. 4, p. 444–445.

Canal
CIVIL ENGINEERING

Canals were built for irrigation purposes by the Assyrians and Egyptians as early as 2000 B.C. In 700 B.C. King Sennacherib (Assyria) built a canal fifty miles long; it was lined with stones, making it waterproof. In more modern times the Suez Canal was built, thereby connecting the Mediterranean Sea and the Red Sea. The French civil engineer Ferdinand de Lesseps completed the Suez Canal in 1869. He failed to construct the Panama Canal in 1888, which was completed in 1914 under the supervision of Colonel George W. Goethals (U.S.). Many canals depend upon **locks** to allow for the different elevations of the land through which they run.

Additional reading: Clark. *Works of man;* p. 13, 98–104.

Candle
ILLUMINATING ENGINEERING

As a means of crude illumination candles found use dating back to ancient times. Cone-shaped candles were included in the decorations of Egyptian tombs built around 3000 B.C. In the Middle Ages candles were made by dipping the wicks into molds containing animal or vegetable fats. Beeswax was later used for producing a cleaner flame. Automation in candle making was introduced in 1950 in a British factory which extruded candles from solid paraffin wax broken into small pieces.

Additional reading: *Illustrated science and invention encyclopedia;* v. 4, p. 441–443.

Cannon
ORDNANCE

The first known drawing of a cannon dates back to 1326 in a book written for England's Edward III by his tutor, Walter de Milemete. In Italy the Council in Florence was given written permission in 1326 to appoint a person to be responsible for making cannons and cannonballs. Cannons were definitely used in the Battle of Crecy in 1346. The first types were small, made of bronze or brass, but cast iron was first used in 1509 in England. The size of the projectiles has varied from one pound to one ton (the latter for the "Great Mortar of Moscow"). Horse-drawn and shipborne types soon appeared.

Additional reading: *Illustrated science and invention encyclopedia;* v. 4, p. 449–451.

Capacitor. *See* **Electric Condenser.**

Capillary
ANATOMY

In 1660 Marcello Malpighi (Italy) used a microscope to observe capillaries in a frog's lungs. By observing capillaries, the smallest vessel found in the circulatory and lymphatic systems, he provided the final link in the circulatory system. Malpighi named these vessels "capillaries," from the Latin word meaning "hairlike." In 1663 Irish physicist and chemist Robert Boyle injected a dye into the blood stream and traced the blood as it passed through the capillaries from the arteries to the veins. Capillary control was demonstrated in the 1900s by Schack August Steenberg Krogh (Denmark). He was awarded the 1920 Nobel Prize in medicine and physiology for showing their role during muscle function. (*See also* Blood Vessel)

Additional reading: Asimov. *Asimov's new guide to science;* p. 598–599.

Carbine. *See* **Rifle.**

Carbon
CHEMISTRY

Carbon, atomic number 6, is known for the characteristic it has of forming a huge number of compounds, probably more than the total of all other elements except for hydrogen. It exists as diamond, graphite, and amorphous carbon. It has been written of, in the diamond form, as early as 1200 B.C. in Indian documents.

The graphite form was identified by the Swedish chemist Karl Scheele in 1779. Diamond and graphite are widely distributed in naturally occurring deposits, while amorphous carbon is a constituent of coal, timber, petroleum, etc. It is widely used in making coke for the iron and steel industry, in industrial chemicals, and in jewelry. Chemical symbol is C. Carbon-44 is used for dating in archaeological studies.

Additional reading: *Van Nostrand's scientific encyclopedia,* p. 520–523.

Carbon Dating — GEOPHYSICS

A method of determining the age of objects was discovered by the American chemist Willard F. Libby around 1947. He used an isotope of carbon (carbon-14), a radioactive substance, to determine ages of objects up to about 45,000 years old. Using other radioactive substances has since extended the range to permit accurate measurement of years in the billions. (*See also* Universe's Origin)

Additional reading: Asimov. *Asimov's biographical encyclopedia;* p. 828–830.

Carborundum — MATERIALS

For 50 years this man-made abrasive was the hardest known substance, with the exception of diamonds. Invented by Edward Goodrich Acheson (U.S.) in 1891, he soon began its manufacture.

Additional reading: Asimov. *Asimov's biographical encyclopedia;* p. 557–558.

Carburetor. *See* Automobile Carburetor.

Carpet Sweeper — MECHANICAL ENGINEERING

In 1876 the owner of a china shop, Melville Bissell (U.S.), patented the carpet sweeper. It took the form of a cylindrical brush that swept dust into a bin.

Additional reading: *Encyclopedia of inventions;* p. 35.

Cartography. *See* Map Making.

Cash Register — MECHANICAL ENGINEERING

An American cafe owner, James Ritty, invented the first cash register in 1879. It had a clocklike face, with the "minute hand" recording cents and the "hour hand" recording dollars. He later added the figures which popped up to show the amount of the transaction and a roll of paper to record the amount. In the 1930s electromechanical models began to appear, followed by electronic types which use integrated circuits to reduce their size.

Additional reading: *Illustrated science and invention encyclopedia;* v. 4, p. 506–508.

Catalysis — BIOCHEMISTRY

During the 1800s Swedish chemist Jons Jacob Berzelius coined the term "catalysis," from the Greek words meaning "break down," to describe the phenomenon wherein chemical substances tended to move from complex to simple structures.

Additional reading: Asimov. *Asimov's new guide to science;* p. 569–570.

Catapult — MECHANICAL ENGINEERING

The catapult, a device for hurling objects, usually used in warfare, was said to have been invented by Dionysus of Syracuse around 400 B.C. The Roman armies also used them, throwing objects up to 500 yards. In modern times catapults are used for the launching of planes from aircraft carriers. The first aircraft carriers used compressed air catapults, but steam became the source of propulsion in 1949 in a British aircraft carrier and has remained the favorite method in the succeeding years. Steam is obtained from the ship boilers.

Additional reading: *Illustrated science and invention encyclopedia;* v. 4, p. 521–522.

Catharsis — PSYCHOLOGY

The concept of "emotional catharsis" was originated by Aristotle, the Greek philosopher who lived from 384 to 322 B.C. Aristotle referred to a purging of pent-up emotions through action such as weeping. Around 1880 the Austrian physician Josef Breuer was the first to utilize the cathartic process to treat hysteria. His patient was known as "Anna O." (*See also* Hypnosis)

Additional reading: Murray. *A history of Western psychology;* p. 298–299.

Catheter — MEDICINE

The first catheter was built in 1929 by German physician Werner Theodor Otto Forssmann. Built with a thin rubber tube, Forssmann's catheter was designed to enable examination of diseased hearts. In 1936 Dickinson W. Richards (U.S.) and Andre F. Cournand (U.S.) made improvements on Forssmann's device and began to experiment on animals. The first successful cardiac catheterization on a human occurred in 1941 under the supervision of Richards and Cournand. The catheter functioned to measure rate of blood flow, oxygen content, and blood pressure.

Additional reading: Robinson. *The miracle finders;* p. 162–164.

Cathode Rays — NUCLEAR PHYSICS

When Heinrich Geissler, a German glass blower, invented in 1854 a way to produce a good vacuum tube enclosing an electrode, he aided those trying to send an electric discharge through a vacuum. Soon scientists noted a green glow opposite the electrode when discharges were produced. In 1876 the German physicist Eugen Goldstein proposed that the glow was caused by the impact of some sort of radiation, which he called cathode rays. Around 1880 the English physicist Sir William Crookes proved that the glow was caused by negatively charged particles. It was not until 1897 that the English physicist Joseph John Thomson verified this and also proved that the glow was caused by subatomic particles, later named **electrons.** (See also Oscilloscope)

Additional reading: Asimov. *Asimov's biographical encyclopedia;* p. 457–458, 561–563.

CD-ROM. *See* Laser Disk.

Cell — BIOLOGY

The Englishman Robert Hooke was the first to use this term and the first to describe cells when, in 1665, he published "Micrographia," an account of his study of the structure of cork. Hooke also observed live plant cells with the aid of a compound microscope. (*See also* Cell Theory)

Additional reading: Magner. *A history of the life sciences;* p. 163–164.

Cell Division. *See* Mitosis.

Cell Nucleus — CYTOLOGY

In 1831 the Scottish botanist Robert Brown discovered nuclei within cells while using a microscope to study plant tissue. Brown named this cellular component "nucleus," from the Latin word meaning "little nut." Another botanist, Matthias Jakob Schleiden (Germany), suggested the role of the nucleus in the process of cell division in 1838. (*See also* Chromosome; Mitosis)

Additional reading: Asimov. *Asimov's new guide to science;* p. 601.

Cell Theory — BIOLOGY

In 1838 and 1839 respectively, Matthias Jakob Schleiden (Germany) and Theodor Schwann (Germany) independently formulated cell theory, stating that all plants and animals consist of cells. The German pathologist Rudolf Virchow extended their theory in 1860 when he suggested that "all cells arise from cells." Virchow demonstrated that cells from diseased tissue were produced by the division of normal cells, and in doing so, founded cellular pathology.

Additional reading: Magner. *A history of the life sciences;* p. 214–236.

Cellophane
MATERIALS

Cellophane, a film made from cellulose, was discovered in 1908 by Jacques Brandenberger (Switzerland). It became commercially available in 1919 when he devised a machine for continuous production of the film. The development of coated cellophane came in 1927 through the work of William H. Church and Karl E. Prindle (U.S.), broadening its uses to such products as wrapping paper.

Additional reading: *Illustrated science and invention encyclopedia;* v. 4, p. 533.

Celluloid
MATERIALS

One of the first products belonging to what is roughly called the plastics family of materials was celluloid. Originally called Parkesine, it was developed by the English inventor Alexander Parkes around 1850, and was used in a variety of manufactured products. Credit is often given to an American printer, John Hyatt, for inventing it in 1867. It remained the most used plastic for many years, not yielding popularity until 1909, when a new plastic known as **Bakelite** was invented.

Additional reading: Clark. *Works of man;* p. 203–205.

Cement. *See* **Concrete and Cement.**

Central Heating. *See* **Heating System.**

Centrifuge
CHEMISTRY

In 1885 Swedish physician Magnus Gustav Blix suggested that centrifugal force could be used to pack red cells together so that their volume could be measured and their number then estimated. In 1889 Sven Hedin (Sweden) followed up Blix's idea when he described a "hematokrit," now called a "centrifuge," which he later built. Hedin's centrifuge consisted of a large wheel connected to a small wheel which held test tubes. When the wheels were turned, the whirling motion forced the blood cells to the bottom of the tubes and their volume could be measured. The number of cells was then calculated. (*See also* Ultracentrifuge)

Additional reading: Reiser. *Medicine and the reign of technology;* p. 132–133.

Centriole. *See* **Centrosome.**

Centrosome
CYTOLOGY

German cytologist Theodor Boveri discovered and named the "centrosome" in the **cytoplasm** in 1887. He observed that the centrosome appeared to be important to the process of cell division. In 1888 Boveri described the centriole, an organelle that formed the center of the centrosome in most cells. He demonstrated that the centriole was usually found near the **cell nucleus** in interphase cells and at the spindle poles during **mitosis.**

Additional reading: Hughes. *A history of cytology;* p. 69–71.

Ceramics
MATERIALS

The class of materials known as ceramics includes not only the ancient types of pottery or the artistic vases and chinaware of past centuries, but also newly discovered materials known for their ability to withstand high temperatures (as in spacecraft construction) or the heavy strength products used for floor tiles or sewer pipes. Thus ceramics dates from ancient pottery made in 7000 B.C. in Iran to ceramic tile forming the outer skin of spacecraft in the United States in the 1980s. (*See also* Glass; Kiln; Porcelain; Potter's Wheel)

Additional reading: *Illustrated science and invention encyclopedia;* v. 4, p. 541–543.

Cerium
CHEMISTRY

Cerium, atomic number 58, is a silver-gray color when it oxidizes in moist air. Three scientists independently discovered it in 1803; namely, M. H. Klaproth (Germany), Jons Berzelius (Sweden), and Wilhelm Hisinger (Sweden). It is used in glass making and in certain alloys, particularly one known as mischmetal. Chemical symbol is Ce.

Additional reading: *Van Nostrand's scientific encyclopedia;* p. 580.

Cesarean Operation
MEDICINE

The first cesarean operation was probably performed in 1500 by Jakob Nufer (Switzerland), although the first fully documented delivery of a fetus through an incision in the abdomen was recorded in 1610 and performed by Jeremiah Trautman (Germany). By the 1700s, Dutch surgeon Hendrik van Roonhuyze had completed many successfully cesareans in The Netherlands. At the same time, however, the procedure was outlawed in Paris because it was believed that it was too dangerous.

Additional reading: McGrew. *Encyclopedia of medical history;* p. 124, 323.

Cesium
CHEMISTRY

A very soft, silver-white metal, cesium, atomic number 55, was discovered in 1860 by the German chemists Robert Bunsen and Gustav Kirchhoff. It rapidly ignites in air, so it is usually stored in kerosene. It is obtained most commonly by treating various minerals, such as pollucite. Its main use is in devices such as photoelectric cells and video equipment. Chemical symbol is Cs.

Additional reading: *Van Nostrand's scientific encyclopedia;* p. 581–582.

Chain Reaction. *See* **Nuclear Fission; Nuclear Reactor.**

Chandrasekhar Mass. *See* **Supernova.**

Character Recognition. *See* **Pattern Recognition.**

Chemical Analysis
CHEMISTRY

The important process known as chemical analysis began only after centuries of superstition and ignorance. The change to scientific methods received an impetus with the work of an Irish physicist and chemist, Robert Boyle, who published a major work in 1661. He was the first to use the term "analysis" in the current chemical sense and also initiated the use of acid-base indicators.

Additional reading: Downs. *Landmarks in science;* chap. 28.

Chemical Elements
CHEMISTRY

The first known scientist to state that a chemical element is a substance which cannot be further broken down into smaller substances was the Irish physicist and chemist Robert Boyle, who wrote on this topic in 1661. He transformed alchemy into chemistry, and was the first to use the term "analysis" in the modern sense. (*See also* names of specific elements; Periodic Table of Elements)

Additional reading: Downs. *Landmarks in science;* chap. 28.

Chemical Equilibrium. *See* **Chemical Reactions.**

Chemical Formula
CHEMISTRY

In the 1850s the field of chemistry was in a rather chaotic condition in regard to methods for writing formulas for compounds. The problem was a lack of general agreement on the relative atomic weights of the elements, causing disagreements about the elementary makeup of various compounds. As an example, there were 19 different chemical formulas for a simple compound like acetic acid in use among various groups of chemists. At an international conference held in 1860 the Italian chemist Stanislaus Cannizzaro convinced chemists of the value of using atomic weights as determined by the Swedish chemist Jons Berzelius in 1828. Soon there was general agreement on the matter.

Additional reading: Asimov. *Asimov's biographical encyclopedia;* p. 439–440.

Chemical Oceanography — OCEANOGRAPHY

Chemical oceanography is the application of chemical techniques, laws, and principles to oceanography. Robert Boyle (Ireland) has been referred to as the father of modern chemical oceanography. One of his accomplishments was the development of the silver nitrate test for seawater around 1680. The uniformity of the composition of seawater was emphasized by M. F. Maury (U.S.) in 1855. Attempts to determine the dissolved oxygen content of seawater were inadequate until the German chemist Clemens Winkler devised a titrimetric method in 1888. The first report of radioactivity in seawater was made by R. J. Strutt (England) in 1906. An early review of the extant knowledge on the degree to which various elements occur in the oceans was made in 1932 by D. W. Thompson (Scotland).

Additional reading: Fairbridge. *Encyclopedia of oceanography;* p. 187–190.

Chemical Reaction — PHYSICAL CHEMISTRY

The general principles of physical chemistry have been largely based upon the work of the American physicist J. Willard Gibbs, publication of which began in 1876. His writings on thermodynamics included emphasis on chemical equilibrium, particularly between different phases (liquids, solids, gases) of chemical systems, leading to his phase rule. A huge number of papers based on the work of Gibbs were subsequently written.

Additional reading: Asimov. *Asimov's biographical encyclopedia;* p. 485–486.

Chemical Structure — CHEMISTRY

The knowledge of **valence** led the German chemist Friedrich Kekule to propose in 1858 a system for showing graphically the way in which the elements in a compound were attached to each other. It had already been realized that the properties of compounds were affected by their structure, such as the obvious case of **isomers,** so a way to show the structure was badly needed. About the same time the Scottish chemist Archibald Cooper made a similar proposal, adding the suggestion that the chemical bonds between atoms be shown with a dash or dotted line. Both these ideas were accepted.

Additional reading: Asimov. *Asimov's new guide to science;* p. 511–512.

Chemical Symbol — CHEMISTRY

The awkwardness of having to refer to chemicals by their full names was eliminated in 1813 when the Swedish chemist Jons Berzelius proposed that elements be referred to by one or two letters representing abbreviations of their Latin names. Thus, Fe stood for iron (ferrum in Latin) and C for carbon. In time this system was universally accepted.

Additional reading: Asimov. *Asimov's new guide to science;* p. 509.

Chiropractic — MEDICINE

In 1895 Daniel David Palmer (U.S.), established a method of treating physical ailments that became known as "chiropractic," when he restored a man's hearing by adjusting his vertebra articulation.

Additional reading: McGrew. *Encyclopedia of medical history;* p. 158–159.

Chloramphenicol — MICROBIOLOGY

In 1943 American botanist Paul Burckholder isolated an antibiotic-producing actinomycete which he named Streptomyces venezuelae. A culture of Burckholder's actinomycete was sent to John Ehrlich (U.S.) and Quentin Bartz (U.S.) and in 1947 they in turn isolated chloramphenicol, so named because of the **chlorine** found in its chemical structure. Also known as chloromycetin, chloramphenicol was the first broad-spectrum **antibiotic** to be discovered. By 1949 Harold Crooks (U.S.), Loren Long (U.S.), John Controulis (U.S.), and Mildred Rebstock (U.S.) had synthesized it and annual sales exceeded 9 million dollars.

Additional reading: Sneader. *Drug discovery;* p. 324–326.

Chlorine — CHEMISTRY

Chlorine, atomic number 17, is, in the gaseous form, a pale greenish yellow gas, having a marked odor and a poisonous effect. It was discovered by the Swedish chemist Karl W. Scheele in 1774 and was confirmed as an element in 1810 by the English chemist Sir Humphry Davy. Its most plentiful form is the common table salt, sodium chloride. Most chlorine is produced by electrolysis of sodium chloride brine. The greatest use is in the production or organic compounds, principally polyvinylchloride. Chemical symbol is Cl.

Additional reading: *Van Nostrand's scientific encyclopedia;* p. 636–638.

Chlorodiazepoxide. *See* **Librium.**

Chloroform — PHYSIOLOGY

Chloroform, a colorless liquid used as an anesthetic, was independently discovered in 1831 by Samuel Guthrie (U.S.), Justus von Liebig (Germany), and Eugene Soubeiran (France). Scottish obstetrician Sir James Young Simpson was the first to use chloroform as an anesthetic when, in 1847, he administered it to a woman during childbirth. Within the same year John Snow (England) adopted its use and also devised a new method of administration, using an apparatus that determined the exact quantity to be applied to the patient. (*See also* Anesthesia)

Additional reading: Sneader. *Drug discovery;* p. 15–22.

Chloromycetin. *See* **Chloramphenicol.**

Chlorophyll — BIOCHEMISTRY

In 1817 French chemists Pierre Joseph Pelletier and Joseph Bienaime Caventou isolated the green compound that gives plants their green color, naming it "chlorophyll" from the Greek words meaning "green leaf." German botanist Julius von Sachs proved that chlorophyll was formed in and confined to certain bodies within the cell named "chloroplasts." Sachs published his research findings in 1865. In 1906 Richard Willstatter (Germany) identified magnesium as the central component of chlorophyll. He was awarded the 1915 Nobel Prize in chemistry for this and other work on plant pigments. German chemist Hans Fischer worked out the complete structure of the chlorophyll molecule during the 1930s. (*See also* Photosynthesis)

Additional reading: Asimov. *Asimov's new guide to science;* p. 590–594.

Chloroplast. *See* **Chlorophyll.**

Chlorothiazide — PHARMACOLOGY

Chlorothiazide was created in 1955 by American chemist Frederick C. Novello. It was an innovative drug that worked as a diuretic to rid the body of excess fluid while also functioning as an antihypertensive. The drug was introduced commercially in 1958 as Diuril and was used to treat congestive heart failure as well as hypertension.

Additional reading: Robinson. *The miracle finders;* p. 184–185.

Chlorpromazine — PHARMACOLOGY

French chemist Paul Charpentier developed chlorpromazine in the 1940s for the Rhone-Poulenc Laboratories. Soon thereafter, French neurosurgeon Henri Laborit discovered the usefulness of its sedating antihistamine attributes for calming surgery patients just prior to the application of anesthesia. Between 1952 and 1955 French psychiatrists Jean Delay and Pierre Deniker found that chlorpromazine calmed manic-depressive patients in addition to working well with schizophrenics. The drug was widely used in the United States by the late 1950s.

Additional reading: "20 discoveries that changed our lives"; p. 141–142.

Chorda Tympani PHYSIOLOGY

In 1843 Claude Bernard (France) discovered the chorda tympani, a fragment of the facial nerve. Bernard showed that it was important for taste and the secretion of saliva.

Additional reading: McGrew. *Encyclopedia of medical history;* p. 266.

Chromatin BIOCHEMISTRY

Derived from the Greek term "chroma," meaning color, this word was coined by Walther Flemming, a German anatomist, upon his discovery in 1879 that threadlike structures within cells are particularly visible when dyes used by cytologists are applied during cell division. These structures make up the deoxyribonucleoprotein complex of which chromosomes largely consist.

Additional reading: Gribbin. *In search of the double helix;* p. 41.

Chromium CHEMISTRY

A hard, blue-white metal, chromium, atomic number 24, is never found in the free state in nature. It is usually obtained from the mineral chromite. It was discovered by the French chemist Nicolas Vauquelin in 1797. The main uses are for decorative electroplating, for hardening certain types of steel, and in the manufacture of numerous compounds. Chemical symbol is Cr.

Additional reading: *Van Nostrand's scientific encyclopedia;* p. 649–652.

Chromosome CYTOLOGY

Walther Flemming (Germany) suggested in 1879 that the filaments found within a **cell nucleus** that pick up dye stain were carriers of hereditary factors. In 1888 Heinrich Wilhelm Gottfried von Waldeyer-Hartz (Germany) coined the term "chromosome" to describe these filaments. In 1887 Edouard Joseph Beneden (Belgium) showed that the number of chromosomes remained constant in the body's cells and was species specific. Edmund Beecher Wilson (U.S.) was the first to see X and Y chromosomes and saw that they were connected to sex determination during the 1890s. In 1902 Walter Stanborough Sutton (U.S.) showed that chromosomes exist in pairs and asserted that they were most likely hereditary factors.

Additional reading: Dunn. *A short history of genetics;* p. 104–115.

Cinchona. *See* Quinine.

Circuit Breaker and Fuse ELECTRICAL ENGINEERING

To protect electric circuits from carrying excessive currents that might lead to fires, it is common to use either circuit breakers or fuses. The fuse is based on the heating effect of electric currents, as stated in 1841 by the English physicist James Prescott Joule. A fuse is made of a metal that melts at low temperatures, thus breaking circuits when currents become too large. Circuit breakers are also used now, especially for handling large currents. They usually contain springs or electromagnets to break a circuit. They are needed in industrial applications where heavy currents are used. Their use probably dates back to the late 1890s, but they were based on the electric relay, invented by Joseph Henry (U.S.) in 1829.

Additional reading: *Illustrated science and invention encyclopedia;* v. 8, p. 1088–1089; v. 14, p. 1845–1849.

Citric Acid Cycle BIOCHEMISTRY

In 1937 Sir Hans Adolf Krebs (England) uncovered the cycle of energy production in living organisms while studying carbohydrate metabolism. He was able to locate two six-carbon acids, including citric acid, involved in a cycle of reactions and intermediate products, from lactic acid to carbon dioxide and water. This cycle is frequently referred to as the Krebs cycle. Krebs shared the 1953 Nobel Prize in medicine and physiology with Fritz Albert Lipmann (U.S.)

Additional reading: Sourkes. *Nobel prize winners in medicine and physiology 1901–1965;* p. 304–314.

Classification of Minerals. *See* Mineral Classification.

Clock HOROLOGY

Clocks have been in existence since the earliest times, the first types being called water clocks because they measured time by the amount of water flowing from one container to another. These were made in China as early as 3000 B.C. Not until the 1300s did mechanical clocks, operated by weights or springs, begin to appear. The first clock accurate enough for scientific use was the pendulum clock, invented by the Dutch physicist and astronomer Christiaan Huygens in 1656, based on experiments made by the Italian astronomer Galileo Galilei in 1581. Electric motors were later used to operate clocks, but the most accurate type yet invented is the **atomic clock.** (*See also* Watch)

Additional reading: Asimov. *Asimov's biographical encyclopedia;* p. 29, 101, 139.

Clone BIOLOGY

In 1952 Robert Briggs (U.S.) and Thomas J. King (U.S.) successfully transplanted living nuclei from blastula cells to enucleated frogs' eggs. In 1956, Americans T. T. Puck, S. J. Cieciura, and P. I. Marcus successfully grew clones of human cells in vitro.

Additional reading: McKinnell. *Cloning of frogs, mice, and other animals;* p. 4, 29–31, 37–38.

Clothes Dryer. *See* Dryer.

Cloud Chamber NUCLEONICS

The study of subatomic particles has benefitted greatly from the cloud chamber, a device invented in 1895 by the Scottish physicist Charles T. R. Wilson. He had worked with containers of clouds formed from water droplets which had condensed on **ions.** After the discovery of X-rays, Wilson believed ions created by radioactivity would bring more intense cloud formations. He found this to be true, but more important, he found the charged particles left tracks. The tracks could be bent by magnetic fields, showing whether the charge was positive or negative and showing the mass of the particle. This later proved to be useful in identifying unknown particles. Photographs create a permanent record of the tracks.

Additional reading: Asimov. *Asimov's new guide to science;* p. 325.

Cloud Seeding. *See* Weather Modification.

Coal Mining MINING ENGINEERING

There is evidence that coal was used in Wales around 2000 B.C., and the mining of coal began in China around 1000 B.C. By 1300 coal mining in England brought such wide use of coal that its use was banned due to air pollution. By 1550 it was again being mined extensively, and cheap labor in the form of women and children was common. They were used to haul coal by hand up ladders or in tunnels too low even for the pit ponies, which were introduced around 1703. Steam power was begun around 1812, when an engine built by George Stephenson (England) moved coal cars on tracks, although boys were also used until the 1900s. The air pump, invented in 1807 by John Buddle (England), aided ventilation. Around 1830 steam-driven fans were used. Coal-mining tools have since been improved.

Additional reading: Derry. *Short history of technology;* p. 467–474.

Cobalt
CHEMISTRY

Cobalt, atomic number 27, is a silvery-gray metal. Although known to have been used as a coloring for Egyptian and Babylonian pottery over 3,000 years ago, it was not identified as an element until 1735 by Georg Brandt, a Swedish chemist. Cobalt must be recovered from various minerals found in many parts of the world. It is used chiefly in high temperature materials in magnetic products, and in wear-resistant materials. Chemical symbol is Co.

Additional reading: *Van Nostrand's scientific encyclopedia;* p. 702–705.

Cocaine
PHARMACOLOGY

During the early 1880s, Austrian psychiatrist Sigmund Freud began testing the mild analgesic properties attributed to cocaine, an extract of the coca shrub. Upon Freud's suggestion, Austrian-born physician Carl Koller (U.S.) began experimenting with cocaine. Koller moved beyond Freud's applications when, in 1884, he successfully used cocaine as a local anesthetic during an eye operation. This breakthrough marked the beginning of the use of **anesthesia** locally. During the late 1880s American surgeon William Stewart Halstead became one of the first to follow up Koller's work by using cocaine injections as a local anesthetic. Experimenting on himself, however, Halstead soon found himself addicted to cocaine and struggled for several years to free himself of the habit.

Additional reading: Sneader. *Drug discovery;* p. 48–57.

Coefficient of Friction. *See* Friction.

Coenzyme
BIOCHEMISTRY

English biochemist Sir Arthur Harden discovered around 1904 that yeast enzyme consisted of two parts: a large protein molecule and a small molecule that was necessary to the functioning of the **enzyme.** The small, non-protein molecule was called a "coenzyme." German-born chemist Hans Karl Euler-Chelpin (Sweden) worked out the structure of Harden's coenzyme in the 1920s. Harden and Euler-Chelpin shared the 1929 Nobel Prize in chemistry for their research. (*See also* Coenzyme A)

Additional reading: Asimov. *Asimov's new guide to science;* p. 710–713.

Coenzyme A
BIOCHEMISTRY

In 1947 German-born biochemist Fritz Albert Lipmann (U.S.) discovered a compound that controlled the transfer of two-carbon groups from one molecule to another. He called that compound "coenzyme A" and showed that a B vitamin, pantothenic acid, made up part of it. Lipmann shared the 1953 Nobel Prize in medicine and physiology with Sir Hans Adolf Krebs (England). In 1951 German biochemist Feodor Lynen found that coenzyme A is important to the breakdown of fats. He also was the first to isolate acetylcoenzyme A, the combination of coenzyme A and the two-carbon fragment. Lynen shared the 1964 Nobel Prize in medicine and physiology with Konrad Emil Bloch (U.S.) (*See also* Coenzyme)

Additional reading: Sourkes. *Nobel prize winners in medicine and physiology 1901–1965;* p. 304–314, 421–431.

Cog Railway
MECHANICAL ENGINEERING

In order to pull trains up steep inclines an engineer named John Blenkinsop (England) devised a system around 1812 involving an extra toothed rail laid inside the tracks in which a gear attached to the locomotive would fit. This prevented slippage on steep inclines. In 1869 this sort of system was used by Sylvester Marsch (U.S.) in the railway built up Mount Washington in the United States. In 1871 the first cog railway in Europe was built by Nicholas Ruggenbach in Switzerland. In 1885 the Swiss engineer R. (?) Abt invented a system having triple gears for a railway in Germany.

Additional reading: *Illustrated science and invention encyclopedia;* v. 14, p. 1909–1910.

Collective Unconscious
PSYCHOLOGY

In 1912 Carl Gustav Jung, a Swiss psychiatrist, published "Psychology of the Unconscious," in which he divided the **unconscious** into two categories—the personal and the collective unconscious. By "collective unconscious," Jung was referring to a realm of the mind that was transmitted by heredity and whose constructs were symbolized by primordial archetypes found in all societies.

Additional reading: Murray. *A history of Western psychology;* p. 311–314.

Color. *See* Color Photography; Light; Motion Pictures; Television.

Color Photography
GRAPHIC ARTS

An early attempt at color photography was made in 1861 by the English photographer Thomas Sutton, who was aided by the Scottish physicist James Clerk Maxwell. Sutton took these pictures, made positive transparencies of them, and projected them onto a screen. But the process of color photography was demonstrated in 1908 on a practical basis by the French physicist Gabriel Lippmann. However, color photography was not to be commercially successful until 1936. In 1959 the American inventor Edwin H. Land presented a new theory of color photography that simplified the matter. (*See also* Motion Pictures)

Additional reading: Asimov. *Asimov's new guide to science;* p. 433.

Color Television
COMMUNICATIONS

The first system for commercial color television broadcasting was invented by Peter Carl Goldmark (U.S.) in 1940. The first regular color television service began in the United States in 1954. The same year, scientists at Bell Laboratories developed standards for practical color broadcasting, sponsored by the National Television System Committee (NTSC). In 1956 Henri de France developed in France a standard for broadcasting called SECAM (Systeme en Couleurs a Memoire); it has been adopted for use in France, Germany, and some other countries. In 1962 Walter Broch (Germany) invented the PAL system (Phased Alternate Line); it is used in Great Britain and other European countries.

Additional reading: *Illustrated science and invention encyclopedia;* v. 18, p. 2377–2383; v. 22, p. 11–12.

Colter
AGRICULTURE

The Romans introduced the "colter" in approximately 1 B.C. They affixed a sharp blade to a pole in front of the **plowshare** that enabled them to break up the heavy, damp soils found in Britain.

Additional reading: Curwen. *Plough and pasture;* p. 62, 80–84.

Combine Harvester
AGRICULTURE

One of the limitations of the **reaper** was that the stalks still had to be tied into bundles and the grain then separated from the straw. In 1835 the American inventor A. Y. Moore developed a machine that combined harvesting and threshing. It was not practical, but was improved about this time by the Australian inventor H. V. McKay. In 1888 ideas for a self-propelled combine harvester were tried out, but not until 1922 did the Canadian firm of Massey-Harris develop a machine for cutting and threshing which was horse-drawn but had machines for the field operations. In 1935 the Allis-Chalmers firm developed a one-man totally self-propelled combine harvester.

Additional reading: *Illustrated science and invention encyclopedia;* v. 5, p. 611–612.

Combustion
CHEMISTRY

A scientific description of the true nature of combustion, replacing the ancient concept of **phlogiston,** was first made by the French chemist Antoine Lavoisier in 1774. His experiments with **oxygen** led to his discovery of the composition of air and of the law of the conservation of mass. He is said to have laid the foundation of modern chemistry. (*See also* Air Composition; Conservation of Mass)

Additional reading: Downs. *Landmarks in science;* chap. 42.

Comet
ASTRONOMY

Comets, usually appearing as rather hazy objects, seem to come and go without any regularity, a fact which puzzled ancient astronomers. Not until 1473 did any European study them with a telescope. In that year the German astronomer Regiomontanus was perhaps the first to do so. It has recently been confirmed that comets consist of ice and dust, with the latter sometimes being freed by evaporation of the ice and forming a tail, which always points away from the sun. The **solar wind** has been found responsible for this. Some comets, like **Halley's Comet,** have a regular schedule of appearances.

Additional reading: Asimov. *Asimov's new guide to science;* p. 151–153.

Communications Code
COMMUNICATIONS

A standard code used for converting symbols into letters and numbers and vice-versa first achieved widespread use when Samuel B. Morse (U.S.) invented around 1836 a code for telegraphy bearing his name. In 1883 a five-unit code was devised by Carl Gauss and Wilhelm Weber (Germany) for use in communication systems. Their code plus one produced in 1877 by J. M. E. Baudot (France) evolved into an international standard code that was approved by the International Consultative Committee on Telegraphs and Telephones (CCITT), known as the CCITT no. 2 code. Similarly standard codes have evolved since the 1950s for computers to permit communication between different models.

Additional reading: *Illustrated science and invention encyclopedia;* v. 17, p. 2351–2352, 2367.

Communications Satellite
COMMUNICATIONS

As early as 1945 a proposal was made by Arthur C. Clarke (England) that satellites in space could aid communications. It was not until 1960 that the first communications satellite, Echo I, was put into orbit by the United States. It was a passive satellite, simply reflecting signals. It had a 100-foot diameter and consisted of aluminum-coated Mylar. Later active satellites were launched; they have separate antennas for receiving and transmitting. Operating at microwave frequencies, they amplify incoming signals some 10 million times before rebroadcasting them. In 1962 the Telstar satellite was launched by the United States. It contained 3,600 solar cells and could transmit both telephone and television signals.

Additional reading: *Illustrated science and invention encyclopedia;* v. 5, p. 620–623.

Compact Disk. *See* **Laser Disk.**

Comparative Psychology
PSYCHOLOGY

In 1882 Canadian-born George Romanes (England) introduced the term "comparative psychology" in his book *Animal Intelligence.* Romanes attempted to relate psychological development to evolutionary advancement.

Additional reading: Murray. *A history of western psychology;* p. 228–231.

Compressor. *See* **Air Compressor.**

Computer. *See* **Analog Computer; Digital Computer; Microcomputer; Supercomputer.**

Computer-Aided Processes
COMPUTER SCIENCE

In 1947 the California Institute of Technology devised a way to use computers to aid designing of aircraft. Since then there have been many computer applications: computer-aided manufacturing (CAM), computer-aided engineering (CAE), computer-aided instruction (CAI), and computer-aided design (CAD), among others. In 1950 MIT developed the first automatically controlled milling machine, used by manufacturers in cutting metal parts. In 1964 computers were being used to produce maps as well as drawings of buildings. In 1968 MIT and the U. S. Air Force developed a CAD/CAM system that would drive several numerical control machines (which operated machine tools such as lathes). In 1974 the Association for Computing Machinery (U.S.) began to develop standards for these processes.

Additional reading: Scott. *Introduction to interactive computer graphics;* p. 162–184.

Computer Graphics. *See* **Computer-Aided Processes.**

Computer Memory
COMPUTER SCIENCE

The designers of the early digital computers realized that they needed better ways to store data in a memory. Around 1948 it was decided to use the mercury delay storage line, replacing dozens of various tubes in the EDSAC, designed by the English inventor Maurice Wilkes and others. The Mark I computer, later in 1948, designed by two Englishmen, F. C. Williams and Thomas Kilburn, used the electron beam from a cathode ray tube to display data on a screen. In 1951 the Whirlwind computer used a magnetic core memory, which consisted of tiny ceramic rings mounted on wire lattices; the inventor was Jay Forrester (U.S.). Since then silicon chips have been created that store huge amounts of data onto thumbnail-sized pieces. (*See also* Bubble Memory; Microchip)

Additional reading: Ritchie. *The computer pioneers;* p. 172–174, 186–188, 193–196.

Computer Printer
COMPUTER SCIENCE

The output of computers is normally prepared by high-speed printers, sometimes called lineprinters (so called because they can print a line at a time). One of the first models was developed in the United States in 1953 by the Remington Rand Company for the Univac computer, having a speed of 600 lines per minute. In 1957 an IBM printer reached 1,000 lines per minute. Thermal printers followed in 1966, made by Texas Instruments Company, then in 1982 the laser printer developed by IBM was able to print 30 lines per second.

Additional reading: *Illustrated science and invention encyclopedia;* v. 10, p. 1378–1379.

Computer Program
COMPUTER SCIENCE

One of the main improvements sought by those designing early computers was that of including a stored program in the computer. The expanded memory needed was provided in the 1940s by use of magnetic delay storage lines. Design of the first stored program is thought to be by the U.S. engineers of the EDVAC—John Mauchly, J. Presper Eckert, Jr., and Herman Goldstine around 1948, but the first computer built with a program was the EDSAC, appearing in 1949, and designed by the English engineer Maurice Wilkes. The first American-built computer to include a stored program was the SEAC, designed by various engineers for the National Bureau of Standards, which went into service in 1950, beating out the EDVAC by a few months. Now stored programs appear in computers of all sizes.

Additional reading: Ritchie. *The computer pioneers;* p. 174, 176, 179, 180.

Concrete and Cement
MATERIALS

The well-known building material known as concrete is typically a mixture of sand and small stones bound together by a cement, usually a hydraulic cement (one that hardens when mixed with water). Cements made of lime and gypsum were used in Egypt for pyramid building around 2700 B.C., and around 150 B.C. Roman builders used a volcanic ash to produce a hard concrete which would harden under water. In 1824 Joseph Aspden (England) patented a process for making Portland cement, named after stones found on the Isle of Portland. It became a very popular constituent of concrete thereafter. Including steel rods in cement, producing reinforced concrete, was first developed in 1845 by the Frenchman J. L. Lambot. More recent developments include precast concrete and prestressed concrete.

Additional reading: *Illustrated science and invention encyclopedia;* v.4, p. 534; v. 5, p. 636.

Condenser. *See* **Electric Condenser.**

Conditioned Reflex
PSYCHOLOGY

Ivan Petrovich Pavlov, a Russian physiologist, demonstrated the existence of conditioned reflexes in 1907 when he published the details of his work with dogs. Pavlov had successfully conditioned a dog to salivate at the sound of a bell rather than at the provision of actual food.

Additional reading: Murray. *A history of Western psychology;* p. 266–273.

Conservation of Energy
PHYSICS

The indestructibility of matter was established in 1778 (*see* Conservation of Mass), but it was not until nearly 40 years later that the conservation of energy received considerable attention. By coincidence three scientists independently stated their hypotheses on energy. In 1842 Julius R. Mayer, a German physicist, issued a paper on the topic, including the mechanical equivalent of heat. In 1847 the English physicist James P. Joule gave a full description of his experiments. In that same year the German physicist Herman von Helmholtz presented his views with even greater detail, eventually winning general acceptance of the law, which is also called the first law of thermodynamics.

Additional reading: Downs. *Landmarks in science;* chap. 61.

Conservation of Mass
CHEMISTRY

The proof that matter cannot be destroyed (it only changes its chemical state) was provided by the French chemist Antoine Lavoisier, around 1778. This discovery stemmed from his study of the process of **combustion.** (*See also* Energy and Mass)

Additional reading: Downs. *Landmarks in science;* chap. 42.

Conservation of Spin. *See* **Particle Spin.**

Constant Composition (Chemical)
CHEMISTRY

At one time there was uncertainty among chemists whether all samples of the same substance had the same properties. Differences were undoubtedly due to impure samples or poor analysis. In 1799 the French chemist Joseph Louis Proust stated what since became known as the law of constant composition; namely, all samples of the same substance contain the same elements in the same ratio by weight. It is also called the law of definite proportions or Proust's law. For example, pure table salt consists of 39.32% of sodium and 60.68% chlorine, measured in terms of weight. However, the statements were not accepted by all chemists, including Claude Louis Bethollet, one of the leading chemists in France. Not until 1808 did Proust prevail over Bethollet.

Additional reading: Garard. *Invitation to chemistry;* p. 119–120.

Contact Lens
OPHTHALMOLOGY

In 1845 Sir John Frederick Herschel (England) suggested contact lenses as a possible corrective device. German physician Adolf Eugen Fick coined the term "contact lens" in 1887, and had a pair made by a German glassblower. The first practical contact lenses were introduced in 1933 by German ophthalmologist Josef Dallos. (*See also* Eyeglasses)

Additional reading: Gorin. *History of ophthalmology;* p. 209.

Continental Drift
GEOLOGY

Continental drift is a geological process in which the continents are drifting apart. At one time it was held in deep scorn by many geologists since it was first presented by the German geologist Alfred Wegener in 1912. Wegener contended that the continents were once all one land mass which he called Pangaea and that the Atlantic and Indian Oceans did not exist 150 million years ago. Looking at maps, the apparent fit of the eastern coast of South America into the western coast of Africa gives some support to the theory. Much evidence has now accumulated in its support, for example, study of the earth's magnetic field, especially the phenonomon of **magnetic reversal** as well as the widening of the Atlantic Ocean. **Plate tectonics** also supports Wegener.

Additional reading: Asimov. *Asimov's new guide to science;* p. 173–175, 189.

Conveyor
MECHANICAL ENGINEERING

Efficiency in handling materials and objects in manufacturing operations has been aided by belts or rollers for moving the materials along. Some conveyors have no power but depend upon gravity for downhill operations. For level or inclined upward movement, power is required. One of the earliest examples was a canvas belt conveyor invented around 1795 by Oliver Evans (U.S.). Bulk materials like coal can be handled; one such device was invented in 1902 by the English inventor W. C. Blackett, and in 1910 the British firm of Drew and Clydesdale built one for handling bananas.

Additional reading: *Illustrated science and invention encyclopedia;* v. 5, p. 656–657.

Copper
CHEMISTRY

Copper, a soft, yellowish-red metal, atomic number 29, dates back to prehistoric times, having been mined for more than 6,000 years. Its earliest use may have been in Egypt around 4000 B.C. It ranks second only to iron in the amount used annually. It is widely distributed in many parts of the world. Among its many uses are electrical conductors, in a variety of alloys and in many industrial compounds. Chemical symbol is Cu.

Additional reading: *Van Nostrand's scientific encyclopedia;* p. 780–785.

Copper Mining
MINING

Copper was one of the earliest metals used by humans, dating back at least to 4000 B.C. By 2000 B.C., the Bronze Age had begun in Egypt where pure copper was being alloyed with tin to make bronze; small plows and tools were soon made of copper or bronze. Clay molds were common then for making decorative objects as well as tools. The Romans learned around 300 A.D. to obtain gold and silver as by-products from ore being treated for its copper content. During the Middle Ages Swedish and German copper supplies grew in importance; by 1500 German copper mines were using improved methods to produce increased amounts of silver as a by-product.

Additional reading: Derry. *Short history of technology;* p. 7, 116, 117, 125, 134, 140, 487.

Copying. *See* **Duplicating; Photography; Xerography.**

Coriolis Force
METEOROLOGY

The rotation of the earth results in an apparent force acting on moving bodies. This effect was first described by the French mathematician Gaspard de Coriolis in 1835 and later expanded in 1855 by the American scientist William Ferrel. In the northern hemisphere this causes winds blowing around low pressure areas to move in a counterclockwise rotation and clockwise around high pressure areas. (*See also* Ekman Layer)

Additional reading: *Illustrated science and invention encyclopedia;* v. 5, p. 664.

Corn Picker
AGRICULTURE

The first mechanical corn picker was marketed in the United States on a small scale in 1909. Two-row corn pickers and huskers were mounted on tractors equipped with power takeoffs and made available to the American farmer in 1928. Finally, in 1946, the first self-propelled corn picker appeared in the United States.

Additional reading: Schlebecker. *Whereby we thrive;* p. 252.

Cornea Transplant
MEDICINE

The first known cornea transplant was performed in 1905 by Eduard Zirm (Austria). Zirm took sections from the patient's cornea and attached them to the eyes of a blind person. He published reports of this operation in 1906.

Additional reading: Gorin. *History of ophthalmology;* p. 455.

Corrosion. *See* **Stainless Steel.**

Cortisone
BIOCHEMISTRY

During the 1930s Edward C. Kendall (U.S.) and Tadeus Reichstein (Switzerland) found cortisone simultaneously. Kendall was searching for adrenal cortex steroids, and by 1938 had isolated six of them, one of which, "compound E," was later named cortisone. Kendall extracted compound E from the adrenal glands of cattle, but in 1941 Lewis H. Sarett (U.S.) discovered a technique for synthesizing it. In 1948 Kendall and his collaborator Philip S. Hench (U.S.) administered compound E to arthritis patients who, as a result of the treatment, experienced a dramatic reduction in pain. Hench then named the compound "cortisone." Kendall, Reichstein, and Hench shared the 1950 Nobel Prize in medicine and physiology for their work.

Additional reading: Robinson. *The miracle finders;* p. 13–19.

Cosmic Rays
PHYSICS

The mysteries associated with cosmic rays have not lessened since they were discovered in 1911 by the Austrian physicist Victor F. Hess. His work involved sending balloons carrying **electroscopes** for detecting radiation as high as six miles above the earth. He stated the radiation came from outer space, and the name of cosmic rays, proposed in 1925 by the American physicist Robert A. Millikan, was accepted. Cosmic rays have a great amount of energy but scientists have no clue as to their source. It is believed they are accelerated by cosmic magnetic fields.

Additional reading: Asimov. *Asimov's new guide to science;* p. 330–331, 343–345.

Cosmology. *See* **Solar System's Origin; Universe's Origin.**

Cotton Gin
AGRICULTURE

The need for a way to remove cotton seed-heads from the cotton fibers led the American inventor Eli Whitney to invent the cotton gin in 1794. This simple machine brought a great reduction in the costs of producing cotton.

Additional reading: *Illustrated science and invention encyclopedia;* v. 5, p. 670–671.

Cotton Picker
AGRICULTURE

In 1926 the first commercial cotton stripper pulled by mules was marketed in the United States. Single- and double-row, tractor-mounted strippers appeared in 1930. All of these early strippers operated by tearing off the entire ball of cotton, whether or not it was ripe. In 1942 International Harvester introduced the spindle cotton picker which picked only ripe cotton. That machine was in wide use in the United States by the 1960s.

Additional reading: Schlebecker. *Whereby we thrive;* p. 245–246.

Cracking (Oil)
PETROLEUM ENGINEERING

The process of cracking consists of breaking large molecules into smaller ones. It is a vital part of the process of **oil refining.** Catalytic cracking, which consists of applying heat and pressure in the presence of a substance known as a catalyst (which aids the reaction without itself being permanently changed), was developed around 1925 by Emile Houdry (France), but the process was not used in the United States until 1936. Gasoline and diesel fuel were two of the main by-products. Thermal cracking relies on heat and pressure alone, while hydro-cracking adds hydrogen to the process in order to produce more high quality gasoline components.

Additional reading: *Cowles encyclopedia of science;* p. 561.

Crane
BUILDING CONSTRUCTION

Examples of ancient constructions which involved stones and masonry of enormous size prove that some sort of lifting devices had to have been used. Cranes for such heavy lifting were described or pictured in the literature of that era. Accounts written by Vitruvius (Italy) around 30 B.C. described a pair of beams having a V-shaped design with a pulley suspended at the top. Guide ropes held it in place. An even earlier use was described by the Roman historian Polybius in writing of the Siege of Syracuse in 212 B.C. He told of cranes designed by Archimedes used for dropping heavy stones or lead weights on enemy ships approaching close to land.

Additional reading: Landels. *Engineering in the ancient world;* p. 84–98.

Critical Temperature
PHYSICS

Studies of the process of **liquefaction of gases** by the Irish physicist Thomas Andrews in 1869 led him to state that there is a temperature (called the critical temperature) above which a gas cannot be liquefied no matter how great the pressure. The theory relating to this phenomenon was developed around 1873 by the Dutch physicist Johannes Van der Waals.

Additional reading: Asimov. *Asimov's new guide to science;* p. 298.

Cryogenics
CRYOGENICS

Extremely low temperatures (near absolute zero) require special techniques to reach. In 1925 the Dutch physicist Peter Debye used a method in which materials that respond to magnetic forces give off heat in proximity to liquid helium. The system was also studied in 1926 by the American chemist William Giauque. Since then a method using helium 3 and helium 4 was used with success by the German physicist Heinz London in 1962. Physicists have reached within a millionth of a degree of absolute zero.

Additional reading: Asimov. *Asimov's new guide to science;* p. 303–304.

Cryosurgery
MEDICINE

American physician Irving S. Cooper developed the technique of cryosurgery for the purpose of freezing and destroying damaged brain tissue. Cooper tried it on a human for the first time in 1961 to relieve symptoms of Parkinson's disease. The technique also proved to be useful for such purposes as removing cataracts and correcting detached retinas. It is still used today as an alternative to L-dopa for treating Parkinson's patients.

Additional reading: Robinson. *The miracle finders;* p. 84–91.

Crystal Growth
CRYSTALLOGRAPHY

Many solid-state devices rely on crystal growing. The usual method is to prepare a molten solution of the material in which crystals grow, after which the concentrated liquid cools. In an important process for producing **semiconductors,** called crystal pulling, a minute particle (or seed) of the material is dipped into the solution, then pulled out so a long thin single crystal is formed; Jan Czochralski (Poland) developed this method in 1917. In zone refining, invented by William G. Pfann (U.S.) in 1952, a molten zone is moved along the crystal until most of the impurities are concentrated at one end of the crystal. Impurities are sometimes purposely added to the molten mix to increase conductivity. (*See also* Synthetic Diamond)

Additional reading: *Illustrated science and invention encyclopedia;* v. 5, p. 688; v. 15, p. 2075–2077; v. 17, p. 2325–2327.

Crystal Rectifier
SOLID-STATE PHYSICS

The German physicist Karl Braun discovered in 1874 that certain crystals allow current to flow in one direction only, allowing such crystals to act as rectifiers, converting alternating current into direct current. This property made crystals popular beginning around 1910 for use in "crystal sets," which were small radio receivers for laypersons. In 1906 the invention of the electron tube was to eliminate the need for crystals in this regard, but the work of the American physicist William Shockley in 1948 in crystal rectifiers and the **transistor** was to revive interest in solid-state devices for communications.

Additional reading: Asimov. *Asimov's biographical encyclopedia;* p. 527–524, 831–832.

Crystallography
CRYSTALLOGRAPHY

An early investigator of crystals (materials having atoms or molecules arranged in a regular pattern) was Nicolaus Steno (Denmark), who in 1669 stated what is now called the first law of crystallography about the fixed angles of crystal faces. In 1771 Jean Rome de l'Isle (France) described the process of crystalization and its relation to natural science. The beginning of the science of crystallography is said to date from the studies by Rene Just Hauy (France) in 1801. Later a major field was to develop in **X-ray crystallography.** A related field is the study of liquid crystals; early work on them was done in 1934 by John Dreyer (England).

Additional reading: *Illustrated science and invention encyclopedia;* v. 5, p. 684–688.

Cultivator
AGRICULTURE

Machines that prepare for seeding the top layer of soil of ploughed land are called cultivators. Prior to World War I, this was done by hand, but in 1912 an Australian inventor, A. C. Howard, developed a prototype machine, improving it in 1922 with a tractor-powered version. These machines are often called rotary cultivators because they contain a series of revolving blades. Modern models combine cultivating, seeding, and fertilizing.

Additional reading: *Illustrated science and invention encyclopedia;* v. 6, p. 689.

Curie Point
PHYSICS

The French chemist Pierre Curie discovered in 1895 that for every substance there is a temperature at which it loses its magnetism. This temperature is now called the Curie point.

Additional reading: Asimov. *Asimov's new guide to science;* p. 224.

Curium
CHEMISTRY

A radioactive metal of the actinide series, curium, atomic number 96, is obtained only by bombardment of heavy elements. It was first identified in 1944 by the American scientists Glenn T. Seaborg, R. A. James, and A. Ghiorso. One of its chief uses has been for thermoelectric generation of power for unmanned space probes or remote terrestrial locations. Chemical symbol is Cm.

Additional reading: *Van Nostrand's scientific encyclopedia;* p. 836–837.

Cybernetics. *See* Robot.

Cyclic AMP
BIOCHEMISTRY

Earl Wilbur Sutherland, Jr. (U.S.) isolated cyclic adenosine monophosphate in 1956, unlocking an important key to hormone functioning. Sutherland found that **epinephrine,** a hormone acting as the "first messenger" outside the cell, stimulates the synthesis of cyclic AMP, the so-called "second messenger," which proceeds to stimulate further biochemical action inside the cell. Sutherland received the 1971 Nobel Prize for physiology and medicine for his research in this area. (*See also* Hormone)

Additional reading: Crapo. *Hormones: the messengers of life;* p. 11–12.

Cyclotron
NUCLEONICS

A particle accelerator that would move particles in an ever widening, circular path was devised in 1930 by the American physicist Ernest Lawrence. The high speeds of the particles caused them to gain mass, upsetting the synchronization. A newer device, called the synchrocyclotron, corrected this in 1946. It was designed by the American physicist Edwin McMillan.

Additional reading: Asimov. *Asimov's new guide to science;* p. 337–338.

Cystoscope
MEDICINE

German physician Max Nitze and Austrian instrument maker Joseph Leiter invented the cystoscope in 1877 to aid in the visual examination of urinary bladders, ureters, and kidneys.

Additional reading: Davis. *Medicine and its technology;* p. 154–155.

Cytochrome
BIOCHEMISTRY

Russian-born biochemist David Keilin (England) discovered cytochrome in 1924 while he was studying the absorption spectrum of horse botfly muscles. He observed a respiratory **enzyme** within the cells that absorbed oxygen, and called it "cytochrome." Cytochrome was found to be an iron-containing enzyme that usually appeared in **mitochondria.**

Additional reading: *The chemistry of life;* p. 35.

Cytoplasm
CYTOLOGY

The term "cytoplasm" was invented in 1882 by German botanist Eduard Adolf Strasburger to describe the **protoplasm** found outside the **cell nucleus.**

Additional reading: Hughes. *A history of cytology;* p. 112–130.

Cytosine. *See* Nucleic Acid.

Czochralski Method. *See* Crystal Growth.

D

DDT
ECOLOGY

Searching for a chlorine-containing compound that would be an effective insect deterrent yet harmless to plants and animals, Swiss-born chemist Paul Hermann Muller discovered that DDT was just such a compound in 1939. DDT was being produced commercially in the United States by 1942, and was used to kill body lice that spread typhus in Italy by 1944. By the 1960s, however, DDT was found to be a harmful environmental pollutant. Muller received the 1948 Nobel Prize in medicine and physiology for his discovery concerning DDT. (*See also* Pesticide)

Additional reading: Sourkes. *Nobel prize winners in medicine and physiology 1901–1965;* p. 261–265.

Definite Proportions Law. *See* Constant Composition (Chemical).

Dental Chair
DENTISTRY

The first dental chair was built in 1790 by Josiah Flagg (U.S.), who was the first full-time dentist. The first reclining dental chair was built by James Snell (U.S.) in 1832. It was upholstered and furnished with a spirit lamp and a mirror fitted on at an angle allowing light to be reflected into the mouth. M. Waldo Hanchett (U.S.) designed the first dental chair to offer a head rest along with an adjustable height and seat back. Hanchett received a patent for his chair in 1848. In 1868 James Beall Morrison (U.S.) designed the first tilting chair which, as it turned out, failed to gain widespread use. The S.S. White Company brought out the first all-metal dentist's chair which could be raised or lowered in 1871.

Additional reading: Ring. *Dentistry: an illustrated history;* p. 194–195, 255–259.

Dental Drill
DENTISTRY

Development of modern drills was accelerated in 1858 when Charles Merry (U.S.) brought out a simple hand-held drill in which a spiral cable rotated the bur. James Beall Morrison (U.S.) improved on that design by attaching a foot treadle and pulleys, a device for which he received a patent in February 1871. In 1868, George F. Green (U.S.) invented the first electric dental drill for the S.S. White Company. Although this drill was put on the market in 1872, it received little use since most dentists did not have electricity in their offices until the 1890s.

Additional reading: Ring. *Dentistry: an illustrated history;* p. 251.

Dentistry
DENTISTRY

Known as the Father of Dentistry, the Frenchman Pierre Fauchard published *Le chirurgien dentiste; ou, traite des dents, (The Surgeon-Dentist, or, Treatise on the Teeth),* thereby largely initiating the era of modern dentistry. In this and his second edition (1746), he organized what was known about dental anatomy, tooth decay, treatment, and dentures, making the body of knowledge accessible for use by others.

Additional reading: Ring. *Dentistry: an illustrated history;* p. 160–172.

Denture
DENTISTRY

In 1756 the techniques needed to build dentures were greatly advanced when a Prussian dentist, Philip Pfaff, published a detailed description of steps to make wax impressions and cast models for missing teeth. Nicolas Dubois de Chemant (France) was granted a patent by Louis XVI for his work creating and perfecting porcelain teeth in 1789. (*See also* Vulcanite)

Additional reading: Ring. *Dentistry: an illustrated history;* p. 170, 180.

Deoxyribonucleic Acid. *See* **DNA.**

Derrick. *See* **Crane.**

Desalination
CHEMICAL ENGINEERING

The processes of making sea water or brackish water fit for human consumption or for irrigation is known as desalination. Efforts to do this date back to Aristotle (Greece) who described evaporation techniques devised by Greek sailors around 400 B.C. Interest continued in the following centuries. A British patent for distillation was granted in 1869. A still to provide industrial water on a large scale was built around 1930 in the Netherlands Antilles. The first large land-based desalination plant was built in Kuwait in 1949, having a capacity of 5 million gallons per day. Some methods used besides distillation have been flash distillation, vapor compression, solar distillation, reverse osmosis, and electrodialysis. (*See also* Osmosis)

Additional reading: *Illustrated science and invention encyclopedia;* v. 6, p. 728–731.

Detonator
ENGINEERING

A device containing a primary explosive that is used to start the explosion of the high explosive material is called a detonator. Detonation involves rapid burning which initiates the explosion, either by great heat or by the shock wave produced. In 1800, the English chemist Edward Howard discovered fulminate of mercury, which the Swedish chemist Alfred Nobel used in 1864 as a detonator for **nitroglycerine.** Lead azide was developed as a detonator in Germany in 1903 and saw widespread use for several decades. In 1866 the English inventor Sir Frederick Abel produced an electric type detonator which is now commonly used. Impact detonators use a firing pin to hit a priming charge. The greatest use of detonators is in coal mining and tunneling. (*See also* Safety Fuse)

Additional reading: *Illustrated science and invention encyclopedia;* v. 6, p. 736–737.

Deuterium
CHEMISTRY

In the early 1930s scientists began a search for an isotope of hydrogen that was heavier than the ordinary hydrogen atom. In 1932 the American chemist Harold Urey was the first to isolate the isotope, named deuterium. This was followed in 1932 by preparation of a sample of water in which deuterium and normal oxygen atoms were present (heavy water), achieved by the American chemist Gilbert Lewis. In 1940 the work of the French physicists Frederic and Irene Joliot-Curie on a nuclear bomb involved using heavy water as a means of slowing down neutrons. The heavy water had to be smuggled out of Norway and later out of France ahead of the Nazis.

Additional reading: Asimov. *Asimov's biographical encyclopedia;* p. 656–567, 739–740, 771–772.

Dewar Flask
CHEMISTRY

A container for storing liquid gases was developed in 1872 by the Scottish chemist Sir James Dewar. His flask made use of a vacuum jacket around a container to eliminate heat transfer by conduction and convection. The domestic thermos bottle was derived from his flask.

Additional reading: *Illustrated science and invention encyclopedia;* v. 6, p. 777–738.

Diacetylmorphine. *See* **Heroin.**

Dialysis. *See* **Artificial Kidney.**

Diamond
MATERIALS

The fact that diamond is chemically the same as graphite was proved in 1772 by the French chemist Antoine Lavoisier and some colleagues. Since this is a reversible relationship, much effort has been spent to convert graphite to diamond. Once it was possible to reach **high pressures,** scientists at the General Electric Company in 1955 used pressures of 100,000 atmospheres and temperatures of 2500 Centigrade to convert graphite to diamond. Even higher pressures and temperatures have since been used. Called **synthetic diamonds,** they are too small and too impure to use for jewelry but are used as abrasives and cutting tools.

Additional reading: Asimov. *Asimov's new guide to science;* p. 305–308.

Diesel Engine
MECHANICAL ENGINEERING

A desire to develop a more efficient engine than the internal combustion engine led the German engineer Rudolf Diesel to invent the Diesel engine in 1897, following several years of experimentation. His engine relied on compressing the fuel-air mixture to the point it ignited itself, without the need for an external ignition source. For many years its most common use was in ships, trucks, buses, and railway locomotives, but in the 1980s it began to find use in some standard automobiles.

Additional reading: Feldman. *Scientists & inventors,* p. 230–231.

Diethylbarbituric Acid. *See* **Sodium Barbital.**

Differential Analyzer. *See* **Analog Computer.**

Differential Gear. *See* **Automobile Transmission.**

Diffraction. *See* **Light Diffraction.**

Diffusion
PHYSICAL CHEMISTRY

When two substances (in gaseous or liquid form) are put in the same container, their molecules will begin to intermingle. In 1831 the Scottish chemist Thomas Graham was able to show that the rate of diffusion of a gas is inversely proportional to the square root of its molecular weight (Graham's Law). His subsequent studies dealt with the diffusion of different substances.

Additional reading: Asimov. *Asimov's new guide to science;* p. 556.

Digestion
PHYSIOLOGY

Between 1825 and 1832, American surgeon William Beaumont performed experiments on Alexis St. Martin, a Canadian who had a hole leading to his stomach as a result of a wound. In particular, Beaumont studied the digestive process and experimented with such variables as temperature and time. Beaumont was able to show that gastric juice was the active substance involved in digestion. In 1833 he published his findings and helped to found modern physiology.

Additional reading: Bettmann. *A pictorial history of medicine;* p. 245.

Digital Computer
COMPUTER SCIENCE

Probably no single invention has had more influence on society than the computer. The speed and accuracy of the digital computer has far surpassed that of **calculating machines.** Many authorities credit the invention of the first general purpose electronic digital computer to three Americans—John Mauchly, J. Presper Eckert, Jr., and John Brainerd. Completed in 1945, their computer contained 18,000 electron tubes, weighed 30 tons, and stood two stories high. It was named the ENIAC (Electronic Numerical Integrator and Calculator). Later improvements included the stored program, faster speed, and large memories. A major change was the development of small but powerful machines for homes and offices. (*See also* Analog Computer; Microcomputer; Supercomputer)

Additional reading: Ritchie. *The computer pioneers;* p. 146–183.

Digitalis
PHARMACOLOGY

While studying herbal remedies, English physician William Withering found that Digitalis Purpurea, commonly known as foxglove, was effective in treating edema caused by heart failure. In 1785 Withering published his data on digitalis, including information on correct dosages, in "An Account of the Foxglove, and Some of Its Medical Uses: With Practical Remarks on Dropsy, and Other Diseases." Digitalis became widely used as a powerful cardiac stimulant.

Additional reading: Sneader. *Drug discovery;* p. 136–140.

Digitizer
COMPUTER SCIENCE

Digitizers are devices for converting analog quantities to digital form. In many cases digital data are easier to read than analog displays; digital data also can be manipulated more easily by computers. Digitizers came into use as digital computers grew in number. In 1952 a **voltmeter** having a digital display was invented by Andrew Kay (U.S.). Another use is to convert graphic designs into digital format for storage in a computer, as is done in certain **computer-aided processes**. Digital-to-analog devices also are extensively used for military applications.

Additional reading: *Illustrated science and invention encyclopedia;* v. 4, p. 418; v. 6, p. 758–760; v. 14, p. 1882; v. 22, p. 17.

Diode. *See* **Crystal Rectifier; Electron Tube; Light Emitting Diode; Semiconductor.**

Diphtheria Antitoxin
IMMUNOLOGY

Friedrich August Loeffler (Germany) discovered the diphtheria bacillus in 1884. His research was advanced by Emile Roux (France) who, between 1888 and 1891, proved that it was this bacillus that caused diphtheria, paving the way for research aimed at developing an antitoxin. In 1892 Baron Shibasaburo Kitasato (Japan), Emil Adolf von Behring (Germany), and Paul Ehrlich (Germany) produced the diphtheria antitoxin, using antibodies that were produced by animals previously inoculated with it. They found that the inoculation not only prevented the disease, but also helped to cure it in animals already infected. The 1901 Nobel Prize in medicine and physiology was awarded to von Behring for this breakthrough. (*See also* Schick Test)

Additional reading: Ackerknecht. *A short history of medicine;* p. 153, 178, 180, 222.

Direct Current Generator. *See* **Electric Generator.**

Direct Current Motor. *See* **Electric Motor.**

Dissection
ANATOMY

Alcmaeon of Croton, a Greek physician born about 535 B.C., was the first person known to have conducted dissections of the human body. Herophilus, a Greek anatomist born about 320 B.C., was the first to perform public dissections of human bodies. (*See also* Postmortem)

Additional reading: Asimov. *Asimov's new guide to science;* p. 595–596.

Distillation
CHEMICAL ENGINEERING

The process in which molecules of a liquid escape from the surface in the form of a gas, which is then condensed and collected, is called distillation. It dates back to the Greek philosopher Aristotle (around 350 B.C.) who describes treating sea water to make drinking water. Later the Arabs used the process to make flavorings and perfumes by distilling plant juices. By 800 A.D. the Islamic scholar al-Jabiz was using vinegar to produce **acetic acid,** probably the only acid known at that time. Distilled alcohol, used at first in medicine, appeared in Italy in the early twelfth century. Liquors began to be made next, and by the early 1800s large scale distilleries had been built.

Additional reading: *Illustrated science and invention encyclopedia;* v. 6, p. 773–777; v. 21, section 29.

Diving Suit
OCEANOGRAPHY

One of the first devices to aid divers was the diving bell, which was invented perhaps as long ago as 300 B.C. Early diving bells were enclosures without fresh air, but in 1752 the English engineer John Smeaton devised an air pump which he fitted to a diving bell. To allow the diver to move about, the diving suit was invented in 1837 by the German engineer Augustus Siebe. In time diving suits came to have an air line and a telephone cable. In order to provide a supply of air to be carried by the diver, the **aqualung** was developed. Other diving devices include the **bathyscaphe.**

Additional reading: *Encyclopedia of inventions;* p. 92–93.

DNA
BIOCHEMISTRY

Johann Friedrich Miescher, a Swiss physician, first discovered DNA in 1869 while studying the composition of white blood cells. Finding it in the cell nuclei, he named it "nuclein." Thirty years later, in 1899, his student Richard Altmann renamed it **nucleic acid.** The botanist Edward Zacharias showed that chromosomes are composed of nuclein in 1881. Arthur Kornberg (U.S.) synthesized DNA in 1956 and shared the 1959 Nobel Prize in medicine and physiology with Severo Ochoa (U.S.). (*See also* Chromosome; RNA; Transforming Principle)

Additional reading: Judson. *The eighth day of creation;* p. 27–29.

DNA Structure
BIOCHEMISTRY

The discovery that the DNA molecule is a double helix composed of two intertwined polyneucleotide chains with a self-complementary structure was made by James Dewey Watson, an American biochemist, and Francis Harry Compton Crick, an English biochemist, in 1953. Watson and Crick shared the 1962 Nobel Prize in medicine and physiology with Maurice Hugh Frederick Wilkins (New Zealand-Great Britain) who accelerated their findings with his study of DNA using X-ray diffraction methodology.

Additional reading: Watson. *The double helix.*

Doppler Effect
PHYSICS

The change in pitch of a sound as the source moves toward or away from a listener was studied by the Austrian physicist Christian Doppler. In 1842 he worked out a mathematical formula explaining the phenomena. Although he predicted it could apply to the frequency (or color) of light, not until after his death was this proved to be true. It also applies to other radiation, such as microwaves. Astronomers have used the Doppler effect to determine the speed of astronomical bodies or even the dimensions of the universe.

Additional reading: Feldman. *Scientists & inventors;* p. 132–133.

Double Helix. *See* **DNA Structure.**

Double Refraction. *See* **Light Refraction.**

Douglas Sea Scale

OCEANOGRAPHY

In order to make estimates of the degree of roughness of oceans a code for this purpose was devised in 1921 by Admiral H. P. Douglas (England). It allowed for separate codes for swell waves (those which are decaying and were originally generated by distant winds) as well as sea waves (fresh waves generated by local winds). The codes were recommended for general use by the 1929 International Meteorological Conference in Copenhagen. They are closely related to the strength of the wind. (*See also* Beaufort Wind Scale; Petersen Scale)

Additional reading: Fairbridge. *Encyclopedia of oceanography;* p. 786–792.

Drifting Ice Station. *See* **Ice Island.**

Drilling Rig

ENGINEERING

The first patent for a rotary drill was granted in 1844 to the Englishman Robert Beart. The first successful oil well was drilled in Pennsylvania in 1859. A wear-resistant diamond bit was developed in 1863 by the French inventor Rodolphe Leschot, and was later used in Pennsylvania to reach a depth of about 3,000 feet. An offshore drilling platform was patented in 1869 by American inventor Thomas F. Rowland, but it was 1897 before the first working model was developed for use off the California coast.

Additional reading: *Illustrated science and invention encyclopedia;* v. 6, 804–809.

Dry Cleaning

MATERIALS

The use of chemicals to clean fabrics had its beginnings in Paris in 1855 when the French dyeworks owner Jean-Baptiste Jolly discovered that kerosene spilled on a tablecloth left that area cleaner than outside the spill area. He began to provide what he called "dry cleaning" from his factory, to distinguish it from the use of soap and water. Other cleaning fluids subsequently used were trichloroethylene and perchloroethylene.

Additional reading: *Illustrated science and invention encyclopedia;* v. 6, p. 813–816.

Dry Dock

NAVAL ARCHITECTURE

Repairing hulls of ships is best done in dry docks, where workers have complete access to the structures. One of the earliest was constructed to accommodate a large warship built for Philopator of Egypt around 205 B.C.; the ship was powered by 4,000 rowers. The dry dock was built at Alexandria and consisted of a large trench dug in the harbor and lined with squared stones. When the ship entered the dry dock the water was removed by a pump (probably the Archimedean screw pump invented by the Greek Archimedes in Egypt around 200 B.C.). The ship was left to settle onto a series of wooden cradles, holding it upright.

Additional reading: De Camp. *The ancient engineers;* p. 151–153.

Dryer

MECHANICAL ENGINEERING

Devices and methods for drying materials have been in use since primitive times, ranging from simply placing objects in sunlight to modern electronically controlled home dryers. An early version of a mechanized clothes dryer was made in 1799 by the French inventor Pochon. It consisted of a crank-operated container pierced with holes. Around 1930, electric clothes dryers first appeared, later to be rivalled by gas-operated models. Spin dryers, first developed around 1909 for an English ocean liner, used centrifugal force to remove moisture. Hair dryers with electric heating elements appeared on the market in the United States around 1920. By 1951 helmeted models for home use were widely available.

Additional reading: *Illustrated science and invention encyclopedia;* v. 4, p. 539–540.

Duplicating

GRAPHIC ARTS

The process of preparing a master copy from which duplicates can be made on a machine is called duplicating, as distinguished from copying from an original. The first types used a wax sheet on which markings were made, with ink squeezed through the markings onto the copy paper. In 1877 the American inventor Thomas Alva Edison developed such a stencil, improved in 1882 by the Hungarian inventor David Gestetner. The mimeograph was developed by A. B. Dick (U.S.) in 1884. The spirit or hectographic method (the "ditto" process) uses a strong aniline dye which is transferred to a backing sheet when it is typed or written upon. This process was developed in Germany in 1923.

Additional reading: *Illustrated science and invention encyclopedia;* v. 6, p. 816–818.

Dye

TEXTILES

Since the earliest times natural materials have been used to dye textiles. A garment discovered in Egypt had been dyed with indigo, its age estimated as 3000 B.C. It was not until 1856 that the English chemist William H. Perkin discovered a process for making dyes from coal tar, previously an unwanted by-product of gas lighting. By the 1860s he was making eight different dye colors. Today more than 1,000 different types of dyes are produced.

Additional reading: Clark. *Works of man;* p. 195–202.

Dynamite

MATERIALS

In 1866 the Swedish inventor Alfred Nobel accidentally discovered that **nitroglycerine** when combined with the substance kisselguhr, which acted as an absorbent, became easier and safer to handle. He named it dynamite and patented it in Great Britain in 1867. In 1871 he invented a gum-like dynamite later known as "plastic."

Additional reading: *Illustrated science and invention encyclopedia;* v. 12, p. 1596–1597.

Dynamo. *See* **Electric Generator.**

Dynamometer

MECHANICAL ENGINEERING

A dynamometer is a device for measuring the power or energy of a machine, such as the horsepower produced by a motor or the power used in an electrical machine. An absorption type dynamometer absorbs and dissipates the power being measured; the simplest type is the Prony brake, invented by the French engineer Gaspard de Prony in 1821. Other types are the Froude (water) brake, the fan brake, and the electromagnetic brake.

Additional reading: *Illustrated science and invention encyclopedia;* v. 6, p. 827–828.

Dysprosium

CHEMISTRY

Dysprosium, atomic number 66, is a soft, silver-gray metal. It was discovered in 1886 by the French chemist Paul Boisbaudran. It can be extracted from a number of minerals found in many countries. It has been used in nuclear reactors to absorb neutrons and as a component of phosphors for television picture tubes. Chemical symbol is Dy.

Additional reading: *Van Nostrand's scientific encyclopedia;* p. 973.

E

Earth's Formation GEOLOGY

The first major effort to understand the forces that governed the formation of the earth was made by a Scotsman, James Hutton. His two-volume work describing his views was published in 1795. He showed a good grasp of the long time periods involved in geological changes, which was contrary to generally held views at the time. (*See also* Solar System's Origin)

Additional reading: Downs. *Landmarks in science;* chap. 40.

Earth's Magnetism. *See* **Terrestrial Magnetism.**

Earth's Mass. *See* **Planetary Masses.**

Earthquake GEOLOGY

One of the first scientists to study earthquakes was the English physicist, Robert Hooke, who proposed around 1670 that they were caused by the cooling and contracting of the earth. A more thorough analysis was made by the English geologist John Michell, who suggested in 1760 that earthquakes set up wave motions in the earth, the timing of which could locate the center of the earthquake. The English geologist John Milne invented in 1880 the **seismograph** for recording earthquake waves. Around 1906 the American geologist Clarence Dutton devised a way to calculate the velocity of seismic waves. In 1935 the American seismologist Charles Richter created a scale for measuring the intensity of earthquakes. (*See also* Plate Tectonics)

Additional reading: Asimov. *Asimov's new guide to science;* p. 162–165.

ECG. *See* **Electrocardiograph.**

Ecological Niche ECOLOGY

In 1908 American zoologist Joseph Grinnell suggested that competition between species was a force that resulted in their adoption of similar yet separate habitats. Each species fit in its own "ecological niche." English animal ecologist C. E. Elton developed the idea further in 1927 when he said that a niche defined the animal's status in its community, especially its relation to food and enemies.

Additional reading: McIntosh. *The background of ecology;* p. 92.

Ecology BIOLOGY

Aristotle, the Greek philosopher who lived between 384 B.C. and 322 B.C., was the first person to write about ecological phenomena and to introduce ecological concepts to scientific literature. In 1866 the German naturalist Ernst Heinrich Haeckel became the first person to use the term ecology to refer to the study of living organisms in relation to one another and their environment.

Additional reading: Sarton. *A history of science;* p. 565–567.

Ecosystem ECOLOGY

The concept of "ecosystem" was introduced in 1935 by British plant ecologist A. G. Tansley. He defined ecosystem in terms of plant communities, and asserted that it was the entire system, encompassing organisms and physical factors.

Additional reading: McIntosh. *The background of ecology;* p. 193–209.

Edison Effect. *See* **Electron Tube.**

EEG. *See* **Electroencephalograph.**

Egg. *See* **Ovum.**

Einsteinium CHEMISTRY

Einsteinium, atomic number 99, is a radioactive metal obtained by bombardment of heavy elements. It was first identified in 1952 by the American scientists A. Ghiorso, S. G. Thompson, G. H. Higgins, and Glenn T. Seaborg during the examination of samples taken from a thermonuclear explosion set off in the South Pacific. No uses are known for the element at this time. Chemical symbol is Es.

Additional reading: *Van Nostrand's scientific encyclopedia;* p. 1022.

Ekman Layer OCEANOGRAPHY

The Ekman layer represents the topmost layer of the ocean in which the surface current can be represented by a balance between frictional forces and the **Coriolis force.** It also applies to the layer of air about 30 feet above the surface up to the region where the wind is not affected by friction. The effect of water friction was calculated in 1902 by the Swedish oceanographer Vagn Walfrid Ekman. Calculations for the atmosphere were made in 1908 by F. A. Akerblom (Sweden).

Additional reading: Hurlbut. *The planet we live on;* p. 178.

Elasticity PHYSICS

The change in shape, size, or volume of a body in response to an applied force and a return to the original dimensions when the force is removed is called elasticity. Materials differ in their degree of elasticity. The first person to make a scientific study of it was the English physicist Robert Hooke in 1678. What is known as Hooke's Law states that a change in length of a body is proportional to the force applied up to the elastic limit of the material (the point at which elasticity no longer occurs). (*See also* Modulus of Elasticity)

Additional reading: *Illustrated science and invention encyclopedia;* v. 7, p. 843–844.

Electret MATERIALS

An insulating material that will keep an electrostatic charge for several years is called an electret. It is the electrostatic equivalent of a permanent magnet. An electret is usually produced on a film in which one side is positively charged and the other negatively charged. It was invented in 1919 by the Japanese scientist Mototaro Eguchi. Early electrets used carnuba wax and shellac; later ones have used polyethylene, Teflon, and polystyrene.

Additional reading: *Illustrated science and invention encyclopedia;* v. 7, p. 845–846.

Electric Arc Process METALLURGY

A successor to the open hearth process of steelmaking is the electric arc process. It was developed in 1898 by Paul Heroult (France), who used large graphite electrodes, to send an electric current through steel scrap. The process provides precise control of the temperature and composition of the molten metal. Originally designed for producing high-quality steel, over half of the output is carbon steels with the rest stainless steel and special alloy steels. Heroult's work had been aided by the invention in 1878 of a small electric furnace by William Siemens (England).

Additional reading: *Illustrated science and invention encyclopedia;* v. 17; p. 2241.

Electric Battery ELECTRICITY

Although many scientists worked to develop a source of electric current, the "voltaic cell" developed by the Italian physicist Alessandro Volta in 1800 is generally considered the first electric battery. Considerable work had also been done in 1791 by Luigi Galvani, an Italian anatomist, who felt that animal muscles were the source of electricity, a belief Volta disproved. A battery producing a constant current over a considerable period of time was invented by the English chemist John Daniell in 1836. The rechargeable battery, called a storage battery, was invented in 1859 by the French physicist Gaston Plante.

Additional reading: Asimov. *Asimov's biographical encyclopedia;* p. 212, 228–229, 313–314, 468.

Electric Condenser
ELECTRICITY

The storage of an electric charge was found possible by experimentation in the 1700s. It required insulation by glass or air, and the first storage device was created in 1745 by Ewald von Kleist (Germany). The first working model was devised a few months later by Peter van Musschenbroek (Holland) at the University of Leyden. It became known as the Leyden jar, and figured in Benjamin Franklin's famous kite flying experiment. Franklin charged a Leyden jar, proving that lightning was composed of electric charges. Condensers (often called capacitors) have since been miniaturized and became much more rugged in design.

Additional reading: *Illustrated science and invention encyclopedia;* v. 4, p. 454–457.

Electric Discharge Tube
PHYSICS

The passage of electricity through a gas-filled tube was a topic of study for the French physicist Antoine Becquerel in 1857. The voltage applied to the tube ends ionized the gas, which would then give off a fluorescent glow. This was the principle for the operation of **fluorescent lighting,** developed years later.

Additional reading: Feldman. *Scientists & inventors;* p. 210–211.

Electric Fan
ELECTRICAL ENGINEERING

The successful development of central **electric power plants** around 1882 led to a sharp increase in the use of electricity for home applicances. Among them was the electric fan, which began to appear on the market around 1890 in the United States.

Additional reading: Strandh. *History of the machine;* p. 225.

Electric Generator
ELECTRICAL ENGINEERING

Devices that transform mechanical energy into electrical energy are called electric generators. The principle governing them was discovered in 1831 by the English physicist Michael Faraday. Both alternating current and direct current electricity can be generated, depending upon generator design. The driving power can be a turbine (fueled by steam, air, water, or sunshine) or a diesel engine. The first practical alternating current generator was built in 1867 by the Belgian-born inventor Zenobe Gramme, who in 1869 also built a generator for producing direct current. Sizes of generators range from small ones for automobiles to large ones for **electric power plants.**

Additional reading: *Illustrated science and invention encyclopedia;* v. 1, p. 78–79; v. 6, p. 825–826; v. 10, p. 1304.

Electric Heating
MECHANICAL ENGINEERING

One of the first uses of electricity for heating was the electric radiator, patented by the English inventors R. E. Bell Crompton and J. H. Dowsing in 1892. In 1906 the American inventor Albert Marsh devised a heating element for electric heaters, made of a nickel-chrome alloy which could be heated white hot before melting. An English inventor who pioneered in this work early in the 1900s was A. H. Barker. Electric heaters rely on radiant heat emanating from radiators or panels. It has the advantage of being very clean.

Additional reading: *Illustrated science and invention encyclopedia;* v. 9, p. 1187–1191.

Electric Induction. *See* **Automobile Ignition System; Electromagnetism.**

Electric Light
ILLUMINATING ENGINEERING

The first incandescent light bulb able to burn for a reasonable length of time was invented by Thomas Alva Edison (U.S.) in 1879. An earlier working model had been made in 1878 by the British inventor Sir Joseph Swan, but he did not get his lamp into production until 1881. In 1910 William Coolidge (U.S.) developed a heat-resisting metal tungsten filament to replace Edison's scorched cotton thread. Many developments in lighting have come about since those years, including the appearance of a rival to incandescent lamps, namely **fluorescent lighting.**

Additional reading: Feldman. *Scientists & inventors;* p. 206–207.

Electric Motor
ELECTRICAL ENGINEERING

A major accomplishment of the American physicist Joseph Henry was his design for the first practical electric motor, prepared in 1831. He saw that an electric current could be used to turn a wheel just as well as turning a wheel could be used to create an electric current--the **electric generator.** The direct current motor was invented in 1837 by Thomas Davenport (U.S.). The Croatian-born electrical engineer Nikola Tesla (U.S.) worked for a while with Thomas Edison, but they parted because Edison favored direct current machines while Tesla chose alternating current. Tesla received a patent in 1888 for such a motor, selling the rights to George Westinghouse. The theory of alternating current machinery was evolved around 1893 by Charles Steinmetz (U.S.).

Additional reading: Feldman. *Scientists & inventors;* p. 224–225.

Electric Power Plant
ELECTRICAL ENGINEERING

The first power station for distributing electricity was designed by Thomas Alva Edison (U.S.), starting service in 1882 for supplying direct current for city lighting in New York. In 1884 Sir Charles Parsons (England) designed a steam turbine to drive the electric generators. In time power stations usually generated alternating current electricity, rather than direct current. Electric generators are often powered by steam created by burning coal or other fuels. In recent years another source developed was the nuclear power station. Still other power stations used water power (the hydroelectric station), **tidal power** and **gas turbine** power. (*See also* Electric Generator; Electric Transformer; Gas Turbine; Hydroelectric Station; Nuclear Reactor; Tidal Power)

Additional reading: *Illustrated science and invention encyclopedia;* v. 14, p. 1837–1842.

Electric Resistor
ELECTRICITY

In 1827 the German mathematician Georg Ohm stated **Ohm's law,** establishing a definition for electric resistance. Resistors have been made by a variety of techniques, one of the most common being that of binding powdered carbon black with an inert filler and molding the mix into insulating sleeves. For higher precision resistors a thin film of metal is deposited onto a base (usually ceramic). For example, in 1919 F. Kruger (Germany) obtained a patent for depositing gold films, thereby creating resistors with very high resistance (up to 200,000 ohms). **Integrated circuits** are replacing individual resistors in many products.

Additional reading: *Illustrated science and invention encyclopedia;* v. 15, p. 1977–1978.

Electric Stove
ELECTRICAL ENGINEERING

Electric stoves began to appear around 1890, and an electric kitchen was created for a New York restaurant in 1894. However, it was not until around the 1930s that electric stoves began to be found in some American homes.

Additional reading: Strandh. *History of the machine;* p. 224–225.

Electric Transformer
ELECTRICAL ENGINEERING

The experiments of Michael Faraday (England) in 1831 with electrical induction resulted in his discovery of the principle of the transformer, a device for increasing or decreasing the voltage in a circuit. A practical way to transport electricity from power plants efficiently was made possible by the invention in 1888 of a suitable transformer by the Croatian-born Nikola Tesla (U.S.). Converting it to high voltage simplified power and transmission. In 1891 a transformer was able to achieve a transmission voltage of 30,000 volts; it was designed by Michael von Dolivi-

Dobrowolsky (Germany) and Charles Brown (Switzerland). Since then many improvements have been made in the quality of transformers of all sizes.

Additional reading: Asimov. *Asimov's new guide to science;* p. 226, 425–426.

Electric Vehicle
MECHANICAL ENGINEERING

As early as 1837 the Scottish inventor Robert Davidson built an electric carriage that had an electric motor and an iron-zinc battery. An electric bus was tried out in London in 1888, and two electric taxicabs that operated for two years in London were built in 1897 by the English inventor W. C. Bersey. In 1899 the Belgian inventor Camille Jenatzy drove his electric car 66 miles per hour. By 1971 a car built by the Eagle Picher Industries in the United States was driven 150 miles per hour on the Bonneville salt flats. The limited range (up to 100 miles) has been a drawback to the use of these cars, but further experimentation is going on. Currently, small vans for urban areas are being used to some extent.

Additional reading: *Illustrated science and invention encyclopedia;* v. 7, p. 857–861.

Electric Washer. *See* **Washing Machine.**

Electricity
ELECTRICITY

Studies of electricity date back to approximately 600 B.C. when the Greek philosopher Thales noted that amber would attract small objects after being rubbed. Around 1600 William Gilbert, an English physician and physicist, studied electrical attraction and introduced the term "electricity" into the language. Around 1749 Benjamin Franklin (U.S.) discovered that electricity has two states, positive and negative; Franklin also studied and explained the nature of lightning. (*See also* Electron; Lightning Rod)

Additional reading: Downs. *Landmarks in science;* chap. 21, 37.

Electrocardiograph
MEDICINE

In 1903 Dutch physiologist Willem Einthoven invented the first crude electrocardiograph in the form of a string galvanometer. He used it to record the electric potential of a beating heart, and suggested that deviation from normal patterns revealed by these recordings could indicate pathological conditions. By 1906 Einthoven was able to correlate his recordings of peaks and troughs, which he called "electrocardiograms," with various kinds of heart disease. He was awarded the 1924 Nobel Prize in medicine and physiology for this work.

Additional reading: Feldman. *Scientists & inventors;* p. 236–237.

Electroencephalograph
MEDICINE

Hans Berger, a German psychiatrist, invented the first electroencephalograph in 1929 to measure the rhythmical electrical activity of the human brain. By attaching electrodes to the brain, Berger was able to "read" the electrical impulses, the most prominent of which he named "alpha" waves and "beta" waves. The electroencephalograph proved useful in the diagnosis of epilepsy.

Additional reading: Asimov. *Asimov's new guide to science;* p. 833–836.

Electrolysis
PHYSICAL CHEMISTRY

In 1800, two English chemists reversed the process used by Alessandro Volta in 1796 to produce an electric current by chemical means; William Nicholson and Anthony Carlisle used electricity to produce chemical changes; namely, decomposing water into hydrogen and oxygen. In 1807 the English chemist Sir Humphry Davy discovered two elements (**potassium** and **sodium**) by the electrolysis of the elements in a fluid state. However, it was not until the English chemist Michael Faraday discovered in 1832 the basic laws of electrolysis that the process was put on a better footing. Faraday became aware of the **ion** at that time. The process is widely used for the production of certain chemicals and for the refining of metals.

Additional reading: *Illustrated science and invention encyclopedia;* v. 7, p. 863–866.

Electromagnet
ELECTROMAGNETISM

One of the many topics of interest to the American physicist Joseph Henry was the electromagnet. He was not the inventor, that being the English inventor William Sturgeon who devised an electromagnet in 1828 by wrapping bare copper wire around a magnet. In 1829 Henry improved it by using insultated wire, which could be wound with overlapping turns without fear of a short circuit. By 1831 his magnet could lift a one-ton weight. A device that opens and closes a circuit when controlled by an electromagnet is called a relay. Henry made use of relays in devising a crude sort of **telegraph.**

Additional reading: Feldman. *Scientists & inventors;* p. 126–127.

Electromagnetic Waves
ELECTROMAGNETISM

Building on the experiments of Michael Faraday (England) in 1831, the Scottish mathematician James Clerk Maxwell evolved in 1864 a set of four equations which described the nature of electricity and magnetism. He explained the relationship of electric fields to magnetic fields and also predicted the existence of electromagnetic waves or radiation and the velocity with which such waves travel. Although **light** was in time recognized as being electromagnetic in nature, other waves were longer in being discovered. It was not until 1888 that Heinrich Hertz (Germany) discovered **radio** waves. Other types of electromagnetic waves are **infrared radiation, ultraviolet radiation, gamma rays, cosmic rays,** and **X-rays.**

Additional reading: Asimov. *Asimov's new guide to science;* p. 225–226, 380–381.

Electromagnetism
ELECTROMAGNETISM

The relationship between electricity and magnetism was first proved in 1819 when the Danish physicist Hans Oersted showed that an electric current affected a magnetic compass. In 1821 the English physicist and chemist Michael Faraday created electric currents by a moving magnetic field, thus inventing the first **electric generator** and discovering electric induction. The ideas of Faraday were interpreted in mathematical terms by the Scottish physicist James Clerk Maxwell between 1864 and 1873. He created mathematical expressions relating electric and magnetic fields. Maxwell and other scientists also studied **electromagnetic waves.**

Additional reading: Asimov. *Asimov's biographical encyclopedia;* p. 281–282, 315–320, 455.

Electron
NUCLEAR PHYSICS

Radiation discovered in a vacuum tube enclosing metal electrodes was originally named **cathode rays.** However, some scientists thought they were electrically charged particles rather than a form of radiation. In 1897 the English physicist Joseph John Thomson showed they were particles by deflecting them by means of an electromagnetic field. He also determined their approximate weight and finally established them as negatively charged particles. The name "electron," which had been suggested in 1891 by the Irish physicist George J. Stoney, was adopted. To further complicate matters, in 1923 the French physicist Louis de Broglie proved that electrons should display the characteristics of waves, with a wavelength inversely proportional to their momentum.

Additional reading: Asimov. *Asimov's new guide to science;* p. 281–282, 403–404, 406–407.

Electron Microscope
ELECTRONICS

A forerunner of today's electron microscope was built in 1932 by the German scientists Ernst Ruska and Max Knoll. It was capable of magnifying an object only 400 times, and was surpassed in 1937 by a machine built by Canadians James Hillier and Albert F. Prebus that magnified up to 7,000 times. In 1939, Vladimir Kosma Zworykin (U.S.) made modifications to the German instrument making it possible for biologists to study viruses and protein molecules for the first time. The first practical scanning electron microscope was built by a British-American physicist, Albert Victor Crewe, in the 1970s. (*See also* Microscope)

Additional reading: Asimov. *Asimov's new guide to science;* p. 404–406.

Electron Tube — ELECTRONICS

One of the discoveries of Thomas Edison (U.S.), made in 1883 and called the Edison effect, was that current would flow in a metal wire sealed in a vacuum tube near the hot filament. Around 1900 the English physicist Sir Owen Richardson showed that hot metal filaments emit electrons in a vacuum tube. In 1904 the English electrical engineer Sir John Fleming used the Edison effect in making a two-electrode tube (a diode) which could convert alternating current into direct current. In 1907 the American inventor Lee De Forest inserted a third electrode, called the grid, into the diode. The grid allowed a small increase in voltage to make a great increase in the flow of electrons. Thus he created an amplifier, which was to revolutionize the design of electronic devices.

Additional reading: Asimov. *Asimov's new guide to science;* p. 443–445.

Electronic Computer. *See* Analog Computer; Digital Computer.

Electronic Countermeasure — ELECTRONICS

The use of electronic countermeasures involves efforts to block reception of **radio** or **radar** or even television signals by various methods. Electronic countermeasures probably originated during World War II, when the various countries involved would jam public radio broadcasts from enemy countries. Also, the jamming of radar detection systems was used to disguise the presence of aircraft or missiles. "Chaff" or small particles were dropped to confuse enemy radar scanners. In the 1980s, a typical countermeasure device of the United States can analyze a quarter of a million radar pulses a second to decide the location of a radar transmitter scanning an aircraft. The complexity of this sort of equipment is constantly growing.

Additional reading: Friedman. *Advanced technology warfare;* p. 36–41.

Electronic Music. *See* Synthesizer (Music).

Electronic Navigation — NAVIGATION

One of the earliest electronic navigational aids was the radio beacon for aircraft invented in 1928 by Donovan (U.S.). In the 1940s systems known as VOR (very high frequency omniranges) and DME (distance measuring equipment) came into use. In 1947 the United States developed Loran (Long Range Air Navigation) and Great Britain made available its Decca Navigator; both provide long distance navigation for both planes and ships. A similar system, developed since the war, is Omega. A special type of microwave device is the Doppler navigator, making use of the **Doppler effect** to measure speed and distance traveled. Around 1960 a self-contained system known as **inertial guidance** found ready use in long-range aircraft and spacecraft.

Additional reading: *Illustrated science and invention encyclopedia;* v. 2, p. 206; v. 12, p. 1574.

Electrophoresis — CHEMISTRY

In 1933 Arne Wilhelm Kaurin Tiselius (Sweden) invented a device which permitted the separation of charged molecules by "electrophoresis." The Tiselius tube, as it became known, made it practical to study **protein** mixtures, particularly blood proteins. Tiselius was awarded the 1948 Nobel Prize in chemistry for his work with electrophoresis. His technique made it possible for Linus Carl Pauling (U.S.) to demonstrate in 1949 that the sickle-cell gene affects the **hemoglobin.**

Additional reading: Asimov. *A short history of biology;* p. 154–158.

Electroplating — METALLURGY

The process of applying a coating of metal to other metal surfaces is called electroplating. An electric current in an electric cell causes metal from the positive plate to pass through a conducting liquid (called an electrolyte) to the negative plate, which is then plated. The process was discovered in 1800 by Johann Ritter (Germany), who electroplated copper. In 1839 patents for electroplating were granted to Germans Carl Jacobi and Werner Siemens. Many types of metals are used for electroplating. The opposite of this process is called electropolishing, which is used to remove irregularities from the surface of plated materials to give them a superior polished surface.

Additional reading: *Illustrated science and invention encyclopedia;* v. 7, p. 885–886; v. 17, p. 2707.

Electropolishing. *See* Electroplating.

Electroscope — ELECTRICITY

The electroscope, a device for detecting electric charges, was invented in 1748 by Jean Antoine Nollet (France). In 1787 Abraham Bennet (England) devised a gold leaf electroscope which gave improved results compared to the previous use of pith balls. In 1791 Alessandro Volta (Italy) introduced metal plates with a layer of insulating lacquer between them.

Additional reading: *Illustrated science and invention encyclopedia;* v. 7, p. 889.

Electroshock Therapy — MEDICINE

Italian physicians Ugo Carletti and Lucio Bini were the first to employ electroshock therapy when, in 1937, they placed electrodes on either side of a schizophrenic's head and sent an electric current through to the patient's brain. Carletti and Bini found the treatment was effective in reducing hallucinations and convulsions. Electroshock therapy was introduced to the United States in 1940 when German-born Lothar B. Kalinowsky (U.S.) demonstrated its effectiveness with depression patients at an American hospital.

Additional reading: Valenstein. *Great and desperate cures;* p. 50–52.

Electrostrictive Effect. *See* Magnetostrictive Effect.

Elementary Particle. *See* Subatomic Particle.

Elevator — MECHANICAL ENGINEERING

Probably the first passenger elevator was invented in 1743 for use by Louis XV of France at Versailles palace. A public elevator was built in 1829 for use in London, and in 1857 an elevator designed by Elisha Graves Otis was installed in an American department store. Hydraulic elevators date back to 1867, when Leon Edoux (France) installed one at an exposition. In 1887 the first electric elevator was built by the German firm of Siemens and Halske for use in tall buildings.

Additional reading: *Cowles encyclopedia of science;* p. 434.

Endoplasmic Reticulum — CYTOLOGY

Belgian-born cytologist Albert Claude (U.S.) discovered the endoplasmic reticulum while studying **mitochondria** with an **electron microscope** in the early 1950s. In 1953 Claude's research partner, Keith Roberts Porter (U.S.), named it "endoplasmic reticulum," meaning "network within the cell fluid." Also known as ER, endoplasmic reticulum provides a structure for holding cell organelles in position. Claude was awarded a share of the 1974 Nobel Prize in medicine and physiology.

Additional reading: Judson. *The eighth day of creation;* p. 266–270.

Endorphins — BIOCHEMISTRY

Endorphins, which are opioid peptides, were sought for several years after the discovery of **enkephlins.** In 1976, Chinese-born biochemist Choh Hao Li (U.S.) and David Chung (U.S.) reported discovery of beta-endorphin, the first endorphin to be found and published in the literature.

Additional reading: Davis. *Endorphins.*

Endoscope — MEDICINE

Arnaud designed the first endoscopic lamp used to illuminate the interior of orifices in humans around 1819. He built his instrument with a biconvex lens.

Additional reading: Woglom. *Discoverers for medicine;* p. 84–99.

Energy and Mass — PHYSICS

Part of a 1905 paper by Albert Einstein (U.S.) on the Special Theory of Relativity included a famous equation relating mass to energy, which can be written as $E=mc^2$, when E is energy, m is the mass of a body, and c is the velocity of light. This theory has since been verified, and has merged into one law the separate laws of **conservation of energy** and **conservation of mass.**

Additional reading: Asimov. *Asimov's new guide to science;* p. 400–403.

Energy Conservation. *See* **Conservation of Energy; Energy and Mass.**

Engine. *See* **Aircraft Engine; Diesel Engine; Internal Combustion Engine; Jet Engine; Steam Engine.**

ENIAC. *See* **Digital Computer.**

Enkephlins — BIOCHEMISTRY

Discovered by John Hughes and Hans Kosterlitz of Scotland in 1974, enkephlins are a type of **opioid receptor** produced in vertebrate brains, and are peptides important to the regulation of pain pathways leading from peripheral nerves to the brain. Their discovery introduced the neuropeptide revolution.

Additional reading: Crapo. *Hormones: the messengers of life;* p. 166–168.

Ensilage Chopper — AGRICULTURE

The first usable ensilage chopper was built in 1915 and combined the cutting machinery of the **corn picker** with a portable **silage** cutter. An improved ensilage chopper was introduced in 1928. It was pulled by a tractor and functioned to cut the corn, chop it, and deliver the cut forage to a truck. The ensilage was then unloaded onto a conveyor and carried into the silo for storage.

Additional reading: Schlebecker. *Whereby we thrive;* p. 253.

Entomology — ZOOLOGY

Dutch naturalist Jan Swammerdam founded modern entomology in the 1600s. In particular, he collected 3,000 insect species and recorded their anatomy with the aid of microscopes. In 1682 he published *Histoire generale des insectes,* in which he reported much of his work. Conclusions from his microscopic dissections of insects were published posthumously in 1737 in *Biblia naturae.*

Additional reading: Magner. *A history of the life sciences;* p. 165–168.

Entropy — THERMODYNAMICS

Entropy can be defined in different ways for different disciplines, such as communications, or mathematics, or statistical mechanics. The last field, which can be used to explain events in **thermodynamics,** defines entropy as a measure of the disorder of a system. In 1854 Rudolph Clausius (Germany) restated a principle of thermodynamics, asserting that in any irreversible change the entropy of a system will increase. This is called the second law of thermodynamics.

Additional reading: *Illustrated science and invention encyclopedia;* v. 18; p. 2401–2402.

Environment. *See* **Pollution Control; Waste and Sewage Disposal.**

Enzyme — BIOCHEMISTRY

The term "enzyme" was introduced in 1878 by Wilhelm Friedrich Kuhne, a German physiologist, to describe substances that bring about chemical reactions associated with living organisms. The first pure enzyme to be crystallized was jackbean urease, prepared by James Batcheller Sumner (U.S.) in 1926. Sumner's work proved that enzymes are proteins and that proteins can act catalytically. (*See also* Coenzyme)

Additional reading: Judson. *The eighth day of creation;* p. 212.

Epinephrine — BIOCHEMISTRY

Also known as adrenaline, epinephrine was the first hormone to be isolated and synthesized. It was identified in 1895 by two Englishmen, George Oliver and Sir Edward Albert Sharpey-Schafer, who found that it is secreted by the adrenal gland and leads to increased blood pressure and heart contractions. In 1897 John Jacob Abel, an American biochemist, was the first to isolate and name it. During the 1950s, Earl Wilbur Sutherland (U.S.) discovered epinephrine's biochemical mechanism of action. This was the first time such information became available for a hormone. (*See also* Cyclic AMP)

Additional reading: Crapo. *Hormones: the messengers of life;* p. 8, 11.

Erbium — CHEMISTRY

Erbium, a soft, silver-gray metal having atomic number 68, was first identified in 1843 by the Swedish chemist Carl Mosander. It is extracted from various minerals found in several countries. It is used for neutron absorption in nuclear reactors, in color television picture tubes, and in high vacuum devices. Chemical symbol is Er.

Additional reading: *Van Nostrand's scientific encyclopedia;* p. 1119–1120.

Erythrocyte — HISTOLOGY

In 1658 Jan Swammerdam (Holland) was the first person to observe and record red blood corpuscles while he was examining aspects of human anatomy microscopically. Anton van Leeuwenhoek (Holland) went on in 1673 to give the first accurate description of red blood cells in the human blood stream. In 1745 Vincenzo Menghini (Italy) discovered that red blood cells contain iron. The first true biconcave form of red blood corpuscles was seen in 1834 by English optician Joseph Jackson Lister. Finally, during the 1850s, French physiologist Claude Bernard showed that red blood corpuscles transport oxygen from the lungs to tissues.

Additional reading: Asimov. *A short history of biology;* p. 28, 124–138.

Erythromycin — MICROBIOLOGY

Americans Robert Bunch and James McGuire of Eli Lilly and Company isolated erythromycin, a crystalline **antibiotic,** from a strain of Streptomyces erythreus in 1952. Erythromycin proved to be a valuable alternative to **penicillin** for patients with allergies to it and for treatment of penicillin-resistant staphylococcal infections.

Additional reading: Sneader. *Drug discovery;* p. 327–328.

Esaki Diode. *See* **Semiconductor.**

Escalator — MECHANICAL ENGINEERING

The earliest moving stairways were installed in 1900 and powered by two endless chains called escalators; they were independently developed in the early 1890s by Jesse Reno (France) and Charles Seeberger (U.S.) in Paris and New York respectively. A more modern type appeared in 1921, using the best features of both inventors. Safety devices are now built into them. Similar in design is the recently developed moving horizontal conveyor used in airports and railway stations.

Additional reading: *Illustrated science and invention encyclopedia;* v. 7, p. 925–927.

Essential Amino Acid. *See* **Amino Acid; Tryptophan.**

Ether DENTISTRY

In 1846 William T. G. Morton, an American dentist, gave the first successful public demonstration of ether used as an anesthetic. Morton administered it to the patient, Gilbert Abbott, as he underwent surgery to remove a tumor from his neck. During the same year, American author and physician Oliver Wendell Holmes suggested the word "anesthesia" to describe the pain killing effect of ether. (*See also* Nitrous Oxide)

Additional reading: Ring. *Dentistry: an illustrated history;* p. 229–237.

Eugenics GENETICS

Sir Francis Galton, an English anthropologist, introduced the term "eugenics" in 1883 when describing his belief that an elite caste should be created by the manipulation of biological breeding. He further urged that defective individuals not reproduce. In the 1950s Hermann Joseph Muller, an American biologist who had access to new information in genetics, asserted that the "science" of eugenics ought to be used to improve the genetic health of the human population. He also supported the establishment of sperm banks to promote use of sperm taken from gifted men.

Additional reading: Medawar. *Aristotle to zoos;* p. 85–92.

Europium CHEMISTRY

A soft, silver-gray metal, europium, atomic number 63, has a high reactivity and must be handled in an inert atmosphere. It was first identified by the French chemist Eugene Demarcay in 1901. It is extracted from several minerals, including bastnasite, found in California. It has been used in control rods for nuclear reactors and in color television picture tubes, as well as other optical devices. Chemical symbol is Eu.

Additional reading: *Van Nostrand's scientific encyclopedia;* p. 1132–1133.

Evolution BIOLOGY

Swiss naturalist Charles Bonnet first used the term "evolution" in the 1700s to describe his belief that periodic catastrophes resulted in increasingly higher life forms. In 1809 French naturalist Jean Baptiste Lamarck became the first biologist of reknown to advance the concept of evolution. His theory rested on the notion of the "inheritance of acquired characteristics." English naturalist Charles Robert Darwin sailed on the H.M.S. Beagle from 1831 to 1836 and collected evidence supporting his theory of natural selection. Darwin published his findings in 1859 in his book entitled *On the origin of species.* In 1972 Niles Eldredge (U.S.) and Stephen Jay Gould (U.S.) proposed the "punctuated equilibrium" model to explain the tempo of major evolutionary change.

Additional reading: Magner. *A history of the life sciences;* p. 341–404.

Exchange Force. *See* **Nuclear Interaction.**

Experimental Physiology PHYSIOLOGY

The Greek physician Galen of Pergamum, who lived from A.D. 130 to 201, established experimental physiology with his studies. He established the function of the recurrent nerve by cutting it and producing loss of voice. He also caused respiration arrest by cutting the medulla.

Additional reading: Ackerknecht. *A short history of medicine;* p. 75–80.

Experimental Psychology PSYCHOLOGY

In 1874 Wilhelm Max Wundt (Germany) published his book entitled *The Principles of Physiological Psychology* and introduced the study of experimental psychology.

Additional reading: Murray. *A history of Western psychology;* p. 171–193.

Expert System COMPUTER SCIENCE

An expert system is usually described as a computer system that solves problems that would normally require handling by a human having expertise. One of the first expert systems was a program that played checkers. Developed in 1947 by Arthur Samuel (U.S.), by 1961 it was playing championship checkers. A chess-playing game was proposed in 1950 by Claude Shannon (U.S.). Early programs were developed thereafter by Manchester University in England and Alex Bernstein (U.S.) in 1956. An early expert system to determine the structure of organic compounds, named DENDRAL, was developed in 1965 by Edward A. Feigenbaum (U.S.). A medical diagnosis system (MYCIN) was developed by Edward Shortliffe (U.S.) in 1972.

Additional reading: Rothfeder. *Minds over matter;* p. 58–63, 69–113.

Explosive MATERIALS

Various types of explosives have been used for many centuries. It is likely the Chinese used potassium nitrate (saltpeter) around 220 B.C. for fireworks. The mixture of saltpeter, sulfur, and charcoal produced a black powder which was first used in England around 1320. Gunpowder was replaced in the nineteenth century by various nitro compounds. Nitrocellulose was invented in 1845 by the German chemist Christian Schonbein, and improved by Sir Frederick Abel in 1875. The development of **nitrogylcerine** in 1846 and **dynamite** in 1866 were important. In recent years ammonium nitrate has been used in a variety of explosives, such as TNT (developed around 1915) and cyclonite or RDX (developed by the German chemist Henning in 1899). (*See also* Detonator)

Additional reading: *Illustrated science and invention encyclopedia;* v. 7, p. 932–935.

Exponent MATHEMATICS

An important work by the French mathematician Rene Descartes published in 1637 included the system of exponents as well as the square root sign, both indispensable to mathematics.

Additional reading: Newman. *World of mathematics;* v. 1, p. 26.

Extrovert-Introvert PSYCHOLOGY

The Swiss psychiatrist Carl Gustav Jung developed and popularized the concepts of "extrovert" and "introvert" in his 1921 publication *Psychological Types.* By extrovert, Jung meant those persons whose psychic energies were focused on outside people and objects. Introverts, on the other hand, were persons whose energies tended to be directed inward, toward their own beliefs and thoughts.

Additional reading: Murray. *A history of Western psychology;* p. 312.

Eyeglasses OPHTHALMOLOGY

Three different people are given credit for inventing the first pair of eyeglasses to correct for farsightedness. In the 1200s, the English scholar Roger Bacon suggested the use of spectacles. According to an inscription over his grave, Salvino d'Armato degli Armati (Italy) invented eyeglasses in 1285. Finally, a Dominican monk, Alessandro della Spina (Italy), was said to have invented eyeglasses in 1299. Nicholas of Cusa (Germany) invented the first eyeglasses to correct for nearsightedness in 1451. Benjamin Franklin (U.S.) designed the first bifocals in 1760. The first eyeglasses to correct for astigmatism were constructed by Sir George Biddle Airy (England) in 1825.

Additional reading: Woglom. *Discoverers for medicine;* p. 104–120.

F

Facsimile Transmission COMMUNICATIONS

The transmission of electrical impulses obtained from the electronic scanning of a document, subsequently converted to print by a second machine, is called facsimile transmission. The first model was invented by the Scottish inventor Alexander Bain in 1842, using a pendulum to create images on a shellac-coated piece of metal. One modern version called Telefax uses telephone lines, but this limits the quality of the transmitted signals. Several choices exist for preparing the copy, including photographic, thermal, and electrolytic techniques. A radio link can also be used, such as transmission of photographs back to earth from **satellites**. Color transmission is being developed for eventual use.

Additional reading: *Illustrated science and invention encyclopedia;* v. 7, p. 948–949.

False Teeth. *See* **Denture.**

Fan. *See* **Electric Fan.**

Feedback System. *See* **Servomechanism.**

Fermentation MICROBIOLOGY

In 1680 Johann Joachim Becher (Germany) showed that sugar was essential for fermentation to occur. Louis Pasteur, the French chemist, determined in 1856 that alcohol was produced during fermentation because yeast cells were present. He also found that if certain rodlike organisms were present during fermentation, the wine or beer soured and lactic acid was produced. In addition, Pasteur showed that the fermentation process does not require oxygen, but does require **yeast,** a living organism. In 1897 Eduard Buchner (Germany) determined that a chemical known as zymase caused alcoholic fermentation.

Additional reading: Magner. *A history of the life sciences;* p. 237–256.

Fermi-Dirac Statistics. *See* **Fermion.**

Fermion NUCLEAR PHYSICS

Most subatomic particles can have one of two kinds of **particle spin.** Those whose spin can be expressed in half-numbers are called fermions. They obey rules worked out in 1926 by the American physicist Enrico Fermi and the English physicist Paul Dirac, known as Fermi-Dirac statistics. Those with the other type of spin are called **bosons.**

Additional reading: Asimov. *Asimov's new guide to science;* p. 342.

Fermium CHEMISTRY

Fermium, a radioactive material of the actinide series, atomic number 100, was discovered in 1952 during an examination of debris from the explosion of an atomic weapon in a test held in the South Pacific. The team of American chemists that made the discovery included S. G. Thompson, A. Ghiorso, G. H. Higgins, and Glenn T. Seaborg. Fermium can be produced by treatment of certain heavy elements. No known uses exist at this time. Chemical symbol is Fm.

Additional reading: *Van Nostrand's scientific encyclopedia;* p. 1164.

Fertilization PHYSIOLOGY

Two biologists working independently made simultaneous observations on the important role the cell nucleus has in fertilization, thus adding new understanding to the process. In 1875, Oskar Hertwig (Germany), using a microscope to study cells extracted from sea urchins, was the first person to observe a male germ cell, the spermatozoon, unite with a female germ cell, the egg. At the same time, Hermann Fol, a Swiss biologist, made similar observations while studying starfish and actually saw the spermatozoon penetrate the egg. Thus, they determined that the nucleus of the **sperm** fuses with the egg nucleus.

Additional reading: Portugal. *A century of DNA;* p. 43–45.

Fertilizer AGRICULTURE

Jean Baptiste Boussingault (France) studied nitrogen requirements of plants in the 1840s. He saw that some plants grew in nitrogen-free soil and water and felt their nitrogen came from the air, not realizing that nitrogen-fixing bacteria existed on their roots. In the 1850s Justus von Liebig (Germany) followed Boussingault's work, but realized that he needed to add nitrogen to his fertilizers, the first synthetically developed. In 1842 Sir John Bennet Lawes (England) patented a process for producing superphosphate and in 1843 built a factory to do so, initiating the synthetic fertilizer industry. In the early 1900s Fritz Haber (Germany) discovered the chemical synthesis of ammonia, making it possible to fix atmospheric nitrogen and convert ammonia to fertilizer.

Additional reading: Asimov. *A short history of biology;* p. 86–88.

Fiber Optics OPTICS

By means of total internal reflection thin glass fibers can transmit light with little loss of intensity. They have a cylindrical core of material (often plastic or glass) surrounded by a material having a lower index of refraction. They have many uses in industry including communication. They were invented in 1955 by the Indian physicist Narinder S. Kapany. The invention of the **laser** in 1960 and of production techniques for glass fibers in 1970 greatly speeded up the development of fiber optics.

Additional reading: *Illustrated science and invention encyclopedia;* v. 8, p. 969–970; v. 22, Section 20.

Fillings DENTISTRY

When Thomas W. Evans, an American dentist, emigrated to France in the 1840s, he introduced the use of silver amalgam for fillings. He is responsible for other innovations in the field, and eventually became the dentist to Emperor Louis Napoleon.

Additional reading: Ring. *Dentistry: an illustrated history;* p. 237.

Fingerprint ANATOMY

Sir William Herschel (England) devised a workable method of fingerprint identification in India in 1869. In 1885 Sir Francis Galton, a British anthropologist, further established that human fingerprints have unique characteristics. Galton developed a system that allowed for individual identification based on the patterns of arches, loops, and whorls that occur with fingerprints. Sir E. R. Henry (England) helped create a method of classifying and filing fingerprints in 1901.

Additional reading: Feldman. *Scientists & inventors;* p. 215.

Finsen Light BIOPHYSICS

The Finsen light was invented in the 1890s by Danish physician Niels Ryberg Finsen. Finsen used his light for the purpose of curing lupus vulgaris, a skin disease caused by the tubercle bacillus, by **irradiation** with strong shortwave light. Finsen's work led to advances in therapy using **X-rays** and **gamma rays.** He was awarded the 1903 Nobel Prize in medicine and physiology.

Additional reading: Sourkes. *Nobel prize winners in medicine and physiology 1901–1965;* p. 15–19.

Fischer-Tropsch Process. *See* **Synthetic Gasoline.**

Fission. *See* **Nuclear Fission.**

Flip-flop Circuit. *See* **Logic Circuit.**

Fluid Mechanics

FLUID MECHANICS

The study of the properties and behavior of fluids in motion consists of hydrodynamics (for liquids) and aerodynamics (for gases). One of the first engineers to apply research methods to the flow of liquids, including design of screw propellers, was the Irish engineer Osborne Reynolds. In the 1880s he devised a formula for the velocity of turbulent versus non-turbulent flow, the ratio now known as Reynolds' number. A leading scientist in the field of aerodynamics was the Hungarian-born Theodore Von Karman (U.S.). In 1938 he aided the development of devices for **jet-assisted takeoffs.** Besides work in 1943 on rocket-propelled missiles, he did extensive studies of **supersonic flight.**

Additional reading: *Engineers and inventors;* p. 121–122, 143–144.

Fluorescence

PHYSICS

When certain materials are exposed to **ultraviolet radiation,** they give off visible radiation called fluorescence. The source could be either natural sunlight or man-made sources. The term was introduced in 1852 by the British scientist Sir George Stokes after studying the phenomenon. Another scientist examining the process was the French physicist Alexandre Becquerel, whose son Antoine was later to discover **radioactivity** while working on fluorescence. In 1857 the older Becquerel used an **electric discharge tube** to study fluorescence. In the early twentieth century **fluorescent lighting** became a commercial success.

Additional reading: Asimov. *Asimov's biographical encyclopedia;* p. 402–403, 406–407, 539–540.

Fluorescent Lighting

ILLUMINATING ENGINEERING

Experiments by the French physicist Alexandre Becquerel with the **electric discharge tube** in 1857 showed that suitable gases would fluoresce when their atoms were ionized by electrons. It was not until 1910 that the French chemist Georges Claude produced a neon light. Electrons would excite neon atoms, giving off a red glow. In 1939 fluorescent lighting as it is known today was demonstrated. It used mercury vapor to produce fluorescence in the phosphor coating on the inside of the tube. It is a cheaper, cooler lighting device than the incandescent light. (*See also* Electric Light)

Additional reading: Asimov. *Asimov's new guide to science;* p. 431–432.

Fluorine

CHEMISTRY

A pale yellow, poisonous gas, fluorine, atomic number 9, was first identified by the Swedish chemist Karl W. Scheele in 1771, although it was not isolated until 1886 by the French chemist Henri Moissan. It is a very reactive, dangerous gas. Only three minerals contain large enough amounts to be commercially useful. It is used as an industrial chemical, in rocket fuels, and in the production of steel and aluminum. Chemical symbol is F.

Additional reading: *Van Nostrand's scientific encyclopedia;* p. 1238–1240.

Fluoroscope. *See* **X-rays (Medicine).**

Flying Boat. *See* **Seaplane.**

FM Radio. *See* **Frequency Modulation.**

Follicle. *See* **Ovum.**

Food Chain

ECOLOGY

In 1891 German zoologist Karl Gottfried Semper introduced the concept of the "food chain" in his book entitled *Animal Life as Affected by the Natural Conditions of Existence.* Semper observed that the **ecosystem** functioned by maintaining a state of energy flow made possible by passing materials through the steps of eating and being eaten.

Additional reading: McIntosh. *The background of ecology;* p. 72, 90.

Forceps. *See* **Obstetric Forceps.**

Fossil

PALEONTOLOGY

The value of fossils as evidence of events of early geological periods has been unquestioned for more than a century. As early as 1500 Leonardo da Vinci (Italy) was aware of the marine origin of fossils. In 1816 the English geologist William Smith used fossils for the first time in such ways as determining the age and order of succession of processes affecting rocks. The importance of fossils was increased by the work of the Scottish geologist Sir Charles Lyell, chiefly through the publication of a three-volume book appearing first in 1830. He startled the world by estimating the age of some fossil-bearing rocks at 240 million years (still less than half the present estimation). (*See also* Paleoecology)

Additional reading: Downs. *Landmarks in science;* chap. 48, 55.

Foucault Pendulum

PHYSICS

The visible proof of the rotation of the earth was given by an experiment made in 1851 by the French physicist Jean Foucault. He noticed that a pendulum keeps swinging in the same plane even if the pendulum's point of suspension moved. Suspending a large iron ball from the center of the dome in the Pantheon in Paris, he proved to the audience that the pendulum's plane of rotation was apparently moving slowly, thus proving his premise that the earth underneath it was actually moving.

Additional reading: Feldman. *Scientists & inventors;* p. 156–157.

Fountain Pen. *See* **Pen.**

Francium

CHEMISTRY

Francium, a radioactive alkali metal, atomic number 87, was discovered in 1939 by the French chemist Marguerite Perey, a collaborator of Madame Curie. Some **thorium** and **uranium** ores containing francium occur in nature. Bombardment of such ores is used to produce it, as well as through similar processes for other heavy elements. There are no known uses at this time. Chemical symbol is Fr.

Additional reading: *Van Nostrand's scientific encyclopedia;* p. 1267.

Fraunhofer Lines

OPTICS

A maker of excellent lenses and prisms, the German inventor Joseph von Fraunhofer made a chance discovery in 1814 while testing prisms. He noticed a large number of dark lines as the sunlight was divided into colors by his prism. He measured the position and wavelength of most of the lines. Starlight also exhibited this phenomenon. He died a few years later before it was found that the dark lines represented the wavelengths of light absorbed by various elements in the atmospheres of the sun and other stars. **Spectroscopy** has since become an important means of analyzing chemical compounds.

Additional reading: Feldman. *Scientists & inventors;* p. 112–113.

Freezer. *See* **Refrigeration.**

Frequency Modulation

COMMUNICATIONS

The process of modifying the frequency of a constant amplitude carrier wave is called frequency modulation. It is used with both radio and television transmission. It was devised by the American electrical engineer Edwin H. Armstrong in 1924, but the first patents were not granted to him until 1933. FM radio is relatively static-free compared to **amplitude modulation,** but requires more room in the bands of assigned frequencies. FM stations gained greatly in popularity after World War II.

Additional reading: *Illustrated science and invention encyclopedia;* v. 8, p. 1067–1068.

Friction

PHYSICS

The resistance to the motion of bodies due to the roughness of surfaces is a simple definition of friction. There are several types of friction, such as sliding friction and rolling friction. In 1492 the Italian artist Leonardo da Vinci was studying the factors that affected friction, such as the nature of the surfaces and the pressure exerted on the surfaces. He also introduced the concept of the coefficient of friction. In 1699 Guillaume Amontons (France) stated that the amount of friction was independent of the areas involved. Nearly 100 years later Charles Auguste Coulomb (France) noted in 1773 that the coefficient of friction depended upon the materials used. He is said to have created the science of friction.

Additional reading: *Dictionary of scientific biography;* v. 1, p. 138; v. 3, p. 441; v. 8, p. 200, 209–210.

Fuel Cell
MECHANICAL ENGINEERING

A fuel cell is a device that converts chemical energy into electric energy, as does an electric battery. In the fuel cell the reactants (normally hydrogen and oxygen) combine to form water, thereby releasing electric energy. Since the gases can be replenished as needed, the unit could continue to supply electric power more or less indefinitely and with a high level of efficiency, making it well suited for use in spacecraft. In 1838 Sir William Grove (England) discovered that the electrochemical dissociation of water was reversible, and then he made a "gas battery" in which several cells were connected in series. In 1936 Francis Bacon (England) began to develop a fuel cell, which was well developed by 1959 and used in the Apollo moon vehicle in the United States.

Additional reading: *Illustrated science and invention encyclopedia;* v. 8, p. 1075–1077.

Fuel Injection
MECHANICAL ENGINEERING

The delivery of fuel under pressure to **internal combustion engines** is always a requirement of **diesel engines**. Air is compressed in the cylinder, thereby heating it sufficiently to cause a spray of fuel injected into the cylinder to be ignited. No electric spark is involved. The first commercial design was probably made by Robert Bosch (Germany) in 1912. Development of fuel injection systems for **aircraft engines** in World War II led to their use in non-diesel engines in lieu of **automobile carburetors** as a means of improving fuel economy. Recent years have seen the use in cars of electronic fuel injection systems (including microprocessors), further aiding fuel economy and reducing exhaust pollutants.

Additional reading: *Illustrated science and invention encyclopedia;* v. 6, p. 747–752; v. 8, p. 1077–1079; v. 23, p. 204–206.

Furnace. *See* **Blast Furnace; Heating System.**

Fuse. *See* **Circuit Breaker and Fuse.**

Fusion. *See* **Nuclear Fusion.**

G

Gadolinium
CHEMISTRY

A soft, silver-gray metal in the rare earth group, gadolinium, atomic number 64, was first identified in 1880 by J. C. G. Marignac, a Swiss chemist. It is obtained chiefly from monazite and bastnasite, found in various countries. It is used in nuclear reactors, in microwave current components, and in color television picture tubes. Chemical symbol is Gd.

Additional reading: *Van Nostrand's scientific encyclopedia;* p. 1309.

Galaxy
ASTRONOMY

A system or huge collection of stars is called a galaxy. Our galaxy has a diameter of about 100,000 light years and contains about 200 billion stars plus masses of dust and gas, and fields of force. There are probably about 200 billion galaxies in the universe. Some have the appearance of a spiral shape. Formerly the term spiral nebulae was used, but the fact that nebulae are galaxies was first discovered by the astronomer Edwin Hubble in 1924. The first person to realize the Milky Way was a galaxy of about 100 million stars was the German-born astronomer Sir William Herschel (England) in 1785. The spiral nature of our galaxy was determined in 1940 by the astronomer William W. Morgan (U.S.). Recently it has been shown that there are clusters of galaxies.

Additional reading: Buedeler. *The fascinating universe;* p. 230–241.

Gallium
CHEMISTRY

Gallium, which is a white, soft metal, atomic number 31, was first identified in 1875 by the French scientist Lecoq de Boisbaudran. Most production is carried out in conjunction with other refining operations, such as aluminum and zinc production. It is used in such products as semiconductors, solar batteries, and dental amalgams. Chemical symbol is Ga.

Additional reading: *Van Nostrand's scientific encyclopedia;* p. 1320–1321.

Galvanometer
ELECTRICITY

Galvanometers are instruments for measuring very small electric currents. They have a movable element that is deflected by a flow of current. They contain a permanent magnet and an electromagnet, either one of which could be movable. The first one was invented by Hans Christian Oersted (Denmark) in 1820. The first moving-coil type was devised in 1825 by Johann Schweigger (Germany), and in 1828 C. L. Nobilli (Italy) designed an astatic type. A galvanometer with a mirror, making it very sensitive, was invented in 1867 by Lord Kelvin (Scotland). Galvanometers can be converted to **ammeters** and **voltmeters.**

Additional reading: *Illustrated science and invention encyclopedia;* v. 8, p. 1095–1096.

Gamma-ray Astronomy. *See* **X-ray Astronomy.**

Gamma Rays
NUCLEAR PHYSICS

With the discovery in 1896 of **radioactivity** came the realization that besides **alpha particles** and **beta particles** there was a third form of emission. This was found to be a penetrating radiation, and it came to be known as gamma rays. The discovery was made in 1900 by the French physicist Paul Villard; the name came from the British physicist, Ernest Rutherford, in that same year. He was able to isolate them in an experiment in 1904.

Additional reading: Asimov. *Asimov's new guide to science;* p. 317–319.

Gas Laws
PHYSICS

Laws concerning the volume, pressure, and temperature of gases were developed by many scientists. One of the first was the Irish physicist and chemist Robert Boyle. In 1662 he discovered an inverse relationship between the volume and the pressure of gases at a constant temperature. A study of gases with a change in temperature was done in 1787 by the French physicist Jacques Charles, but he did not publish his findings. This remained to be done in 1802 by the French chemist Joseph Gay-Lussac. Further refinements were made by the Dutch physicist Johannes Van Der Waals in 1873. Related work led to the **liquefaction of gases.** (*See also* Kinetic Theory of Gases)

Additional reading: Asimov. *The history of physics;* p. 191–192.

Gas Lighting
ILLUMINATING ENGINEERING

The lighting of buildings and streets, now taken for granted, had its origin in England in the experiments of the engineer William Murdock (Scotland). In 1803 he used gas from burning coal to light a room. In France an engineer, Philippe Lebon, patented in 1799 a lamp which burned gas from wood distillation. Murdock illuminated his entire factory building in 1803, and gas companies soon appeared in all cities in England.

Additional reading: Feldman. *Scientists & inventors;* p. 86–87.

Gas Stove
MECHANICAL ENGINEERING

In 1802, the year of the invention of the coal stove, a gas stove was invented by the German-born Frederick Winsor (England). By the 1840s gas stoves were being built in the United States, and by the 1870s they had become popular in American homes.

Additional reading: *Encyclopedia of inventions;* p. 34.

Gas Turbine
MECHANICAL ENGINEERING

A device that converts the energy of a gas into mechanical energy is called a gas turbine. Normally the energy is used to rotate a shaft. In 1791 John Barber (England) patented a gas turbine, but he never succeeded in producing a working model. The word "turbine" was coined in 1847 by the French engineer Claude Bourdin, and he also proposed building a turbine. By 1901 commercial models were in production in France. Other types of turbines include **steam turbines,** water-powered turbines (such as are used in **hydroelectric stations**), and aircraft turbines (as are used in **jet engines**).

Additional reading: Feldman. *Scientists & inventors;* p. 300–301.

Gasoline. *See* **Synthetic Gasoline.**

Gasoline Engine. *See* **Internal Combustion Engine.**

Gear
MECHANICAL ENGINEERING

Toothed wheels, or gears, were used by the ancients for various devices, such as pulleys and the **water wheel.** The use of gears is mentioned in the writings of Vitruvius (Italy) around 30 B.C. and Hero (Greece) around 50 A.D. The latter writer invented a device called a hodometer which used gears for converting wheel revolutions of a chariot into distance. He was also an early investigator of the nature of the inclined plane and the lever. Around 1492 Leonardo da Vinci (Italy) designed several types of gears, such as helical, conical, and trapezoidal. In 1657 Gaspar Schott (Germany) created a classification system for different types of gear teeth. Gears have continued to be vital parts of all sorts of mechanisms.

Additional reading: *Dictionary of scientific biography;* v. 8, p. 210; v. 12, p. 211; v. 15, p. 519.

Geiger Counter
NUCLEONICS

A measuring device that uses ionization of a gas-filled tube when exposed to charged particles is called a Geiger counter. It was invented by the German physicist Hans Geiger in 1908. In 1913 Geiger and Walther Muller (Germany) improved the earlier model, making it more sensitive and broadening its uses. It is primarily used as a simple way to locate a source of radioactivity. It has a loudspeaker so that an increase in amount of ionization increases the number of audible clicks. More sophisticated detectors are available now.

Additional reading: *Illustrated science and invention encyclopedia;* v. 9, p. 1114–1115.

Gene
GENETICS

Danish botanist Wilhelm Ludwig Johannsen introduced the term "gene," from the Greek word meaning "to give birth to," in 1909 as a replacement for the Mendelian term "factor." In 1915 Americans Thomas Hunt Morgan, Calvin Bridges, Alfred Henry Sturtevant, and Hermann Joseph Muller published "The Mechanism of Mendelian Heredity" in which they asserted that chromosomes contain invisible genes that determine the hereditary traits of offspring. (*See also* Genetics)

Additional reading: Gribbin. *In search of the double helix;* p. 56-82.

Genetic Code
GENETICS

Russian-born George Gamow, an American physicist, suggested in 1954 that nucleic acids act as a "genetic code" during the formation of **enzymes**. His idea was proven correct in 1961 when Marshall Warren Nirenberg (U.S.) used synthetic RNA to serve as **messenger-RNA** and worked out the first item of the genetic code. Nirenberg shared the 1968 Nobel Prize for medicine and physiology with Har Gobind Khorana (U.S.) and Robert William Holley (U.S.). Khorana, independent of Nirenberg, worked out almost the entire genetic code and introduced techniques to compare DNA of known structure with the RNA it would produce, thus showing that the separate nucleotide triplets or letters of the code would not overlap.

Additional reading: Judson. *The eighth day of creation;* p. 225–491.

Genetic Engineering
BIOLOGY

In 1973 Stanley Cohen (U.S.) and Herbert Boyer (U.S.) marked the birth of genetic engineering when they successfully constructed DNA molecules by combining genetic material from two different sources. The United States Supreme Court ruled in 1980 that genetically modified microorganisms can be patented, thereby affirming a patent Cohen and Boyer had filed covering the broad technology involved in genetic engineering. (*See also* Clone)

Additional reading: Nossal. *Reshaping life;* p. 104, 126–127.

Genetic Map
GENETICS

In 1911 Thomas Hunt Morgan, an American geneticist, and his pupil, Alfred Henry Sturtevant (U.S.), drew up the first chromosome maps for fruit flies. Two years later Sturtevant published the first genetic map and developed the principle of mapping gene position on chromosomes by determining the frequency with which crossing over separated the genes. Finally, in 1951 Sturtevant presented a map of the fourth and smallest of the Drosophila chromosomes which he had mapped using his technique. (*See also* Chromosome)

Additional reading: Gribbin. *In search of the double helix;* p. 56–82.

Genetic Recombination
GENETICS

American geneticist Joshua Lederberg and American biochemist Edward Lawrie Tatum demonstrated genetic recombination in bacteria in 1946. They found that different strains of bacteria could be crossed to make their genetic material intermingle. Lederberg and Tatum shared the 1958 Nobel Prize in medicine and physiology with George Wells Beadle (U.S.). Also in 1946, Max Delbruck (U.S.) and Alfred Day Hershey (U.S.), independently of each other, demonstrated genetic recombination of **bacteriophage.** They shared the 1969 Nobel Prize in medicine and physiology with fellow American Salvador Edward Luria.

Additional reading: Sourkes. *Nobel prize winners in medicine and physiology 1901-1965;* p. 352–368.

Genetics
BIOLOGY

The study of genetics began in 1865 when Gregor Mendel (Austria) began experiments in which he crossbred various strains of garden peas. His results led him to believe that an organism carries within it a set of hereditary factors that is transmitted to offspring. Mendel published his findings in 1866 in "Versuch ueber Pflanzenhydriden," a paper that was ignored until 1900 when it was rediscovered independently by Carl Correns (Germany), Hugo de Vries (Holland), and Erich Tschermak (Austria). In 1905 William Bateson, an English biologist, suggested that the study of inheritance be called "genetics." (See also Gene)

Additional reading: Gribbin. *In search of the double helix;* p. 27–55.

Genotype
GENETICS

Danish botanist Wilhelm Ludwig Johannsen coined the term "genotype" in 1909 to refer to the genetic constitution of an organism. (*See also* Phenotype)

Additional reading: Dunn. *A short history of genetics;* p. 88–95.

Geodometer. *See* **Surveying.**

Geometry
MATHEMATICS

Euclid, a Greek mathematician, has been called the father of geometry, primarily because of his textbook *Eleements of Geometry*. Actually many of the theorems in his famous book were developed several hundred years earlier, but he not only codified over 250 years of work but also added important theorems of his own in his text, written around 300 B.C. In the nineteenth century a newer type of geometry was developed, known as **non-Euclidean geometry.** (*See also* Analytic Geometry)

Additional reading: Downs. *Landmarks in science;* chap. 4.

Germ Theory
MEDICINE

Friedrich Gustav Jakob Henle (Germany) suggested in 1840 that disease was caused by microorganisms. The French chemist Louis Pasteur's work with **fermentation** led him also to believe that microbes might be responsible for disease. Finally, in 1865, Pasteur established that germs were at the root of a silkworm disease he was studying, leading him to hypothesize that disease was communicable because germs both caused and spread it. It was in this way that Pasteur developed the germ theory of disease.

Additional reading: Magner. *A history of the life sciences;* p. 253–257.

Germanium
CHEMISTRY

Germanium, a hard, silver-white brittle metal, atomic number 32, was identified in 1886 by the German chemist Clemens Winkler. It is found in small quantities in various minerals, but the primary source is flue dust from the processing of zinc. Its chief use is in the production of **transistors,** once techniques for purifying it by single **crystal growth** and zone refining were perfected. Chemical symbol is Ge.

Additional reading: *Van Nostrand's scientific encyclopedia;* p. 1368–1369.

Gestalt Psychology
PSYCHOLOGY

Max Wertheimer (Germany) and his colleagues Kurt Koffka (Germany) and Wolfgang Kohler (Germany) began the gestalt psychology movement during the 1900s. Wertheimer described what he termed the "phi phenomenon," wherein stationary lights appear to have movement. This assertion marked the beginning of gestalt psychology, an approach which stated that perceptual qualities not existing in parts did exist in a whole.

Additional reading: Murray. *A history of Western psychology;* p. 243–257.

Glacier
GEOLOGY

For many years there were several theories about the cause for large boulders scattered through many countries. Some geologists felt they had been deposited by glaciers while the land was under water. The Scottish geologist, James Hutton, devoted much study of the source of boulders, but his writings in 1795 did not attract as much attention as did the paper delivered in 1837 by the Swiss biologist Louis Agassiz, who devoted years of study to the effects of glaciers on dry land, not in areas under water. Today his views are accepted, including the descriptions he gave of the Ice Age, in which huge ice sheets covered continents for perhaps a million years.

Additional reading: Asimov. *Asimov's biographical encyclopedia;* p. 193–194, 362–363.

Glass
MATERIALS

As early as 4000 B.C. glass was used in Egypt to glaze the surfaces of soapstone beads, while glass was used around 2500 B.C. in imitations of precious stones. Glass containers appeared by 1500 B.C. while glass blowing did not start until 300 B.C. By 200 A.D. the art of glass blowing spread from Syria to Italy, then across Europe. After the Dark Ages glass making resumed in Greece around 1000 and in Venice around 1200, from whence it went as far as England. By around 1300 fine stained glass windows were being made for cathedrals in England and France. Meanwhile Italian stained glass workers emphasized making vessels, both clear and colored. The first bottle-making machine was built by T.(?) Ashley (England) in 1887. (*See also* Lens; Plate Glass)

Additional reading: *Illustrated science and invention encyclopedia;* v. 9, p. 1122–1127.

Glass Eye
OPHTHALMOLOGY

Artificial clay eyes were used by ancient Egyptians, but it was not until the 1600s that John Scultetus (Germany) invented the glass eye. At first, glass eyes were manufactured in Venice, and later in Bohemia and France. Between 1835 and 1933 the glass eyes manufactured in Germany were considered the finest.

Additional reading: McGrew. *Encyclopedia of medical history;* p. 232.

Glider
AEROSPACE ENGINEERING

Probably the first heavier-than-air device for flying was a glider built in 1804 and designed by Sir George Cayley (England). In 1809 he wrote a periodical article on the principles of powered flight, the first ever published on aerodynamics. Another pioneer was the German mechanical engineer Otto Lilienthal who, with his brother Gustav, built and flew what could be called hang gliders. In 1895 Otto flew 1,000 feet, but in 1896 he was killed in a crash of one of his gliders. The Lilienthals made discoveries which were later to aid the Wright brothers in their efforts to build an **aircraft.** They too used gliders before building motor-driven craft.

Additional reading: Clark. *Works of man;* p. 238–244.

Globular Clusters
ASTRONOMY

For several years beginning in 1918 the American astronomer Harlow Shapley studied what are called globular clusters. They consist of up to several million stars in one cloud (with diameters of the cluster of about 100 light years). There are about 100 globular clusters in our **galaxy**.

Additional reading: Asimov. *Asimov's new guide to science;* p. 30–32.

Glutathione
BIOCHEMISTRY

English biochemist Sir Frederick Gowland Hopkins isolated glutathione in 1921 from living tissue, and demonstrated its importance to plant and animal cell oxidative processes.

Additional reading: *The chemistry of life;* p. 35.

Gold
CHEMISTRY

One of the oldest metals known to man, gold, atomic number 79, was being used in jewelry as far back as 4000 B.C. in Mesopotamia. Written records of all countries mention the use of gold. It is yellow, soft, and very malleable. It can be beaten into films less than five millionths of an inch thick. Gold is usually found as the free metal in gravel or scattered in veins of quartz, as well as in sea water. It is primarily used in jewelry as well as in certain electronics products, in dental work, and in some industrial alloys. Chemical symbol is Au.

Additional reading: *Van Nostrand's scientific encyclopedia;* p. 1392–1394.

Gold Mining
MINING

Gold cups from Iraq date back to 3500 B.C.; Egypt had more than 100 gold mines around 2500 B.C. Gold was refined then in porous clay crucibles, with blasts of air to remove impurities. The Romans around 1 A.D. developed a method of using mercury for extracting gold after crushing the ore. During the Middle Ages much effort was wasted on **alchemy**, the attempt to convert base metals into gold. In some areas gold has been obtained by sifting gravel and sand until only the gold is left. In other regions underground mining required techniques much like those of **coal mining.** In the 1500s much gold and silver were mined in South America and Mexico, brought back to Europe by the Spanish. Numerous deposits have since been discovered in many regions.

Additional reading: Derry. *Short history of technology;* p. 116, 259, 491.

Golgi Cell CYTOLOGY

In 1873, when Italian histologist Camillo Golgi introduced the use of silver salts for cellular staining, the "Golgi bodies" and "Golgi complex" were revealed for the first time. The function of these cellular components has never been discovered. Golgi shared the 1906 Nobel Prize in medicine and physiology with Santiago Ramon y Cajal (Spain) for his work. (*See also* Neuron)

Additional reading: Asimov. *Asimov's new guide to science;* p. 816–833.

Graham's Law. *See* **Diffusion; Kinetic Theory of Gases.**

Grain Elevator AGRICULTURE

In 1785 Oliver Evans (U.S.) invented and patented the first elevator designed to handle grain. Evans used an endless chain bucket system to move the grain from bins into a storage elevator. It was then taken in bulk to a mill where another endless chain bucket system moved the grain to the top of the mill. The grain was ground to produce flour and a by-product called "middlings," which could be used as animal feed. In 1843 Joseph Dart (U.S.) adapted the grain elevator to take advantage of steam power. His system made use of the elevator primarily for storage rather than milling. **Steam engines** operated the chain buckets.

Additional reading: Schlebecker. *Whereby we thrive;* p. 133–134.

Gramophone. *See* **Phonograph.**

Gravitation PHYSICS

One of the most important accomplishments of the English scientist Sir Isaac Newton was the law of universal gravitation, which he discovered around 1665. He showed that the force applied to celestial bodies as well as to the earth and that its magnitude was directly proportional to the product of the masses of two bodies and inversely proportional to the square of the distance between their centers. In 1798 an important part of Newton's formula, the gravitational constant, was successfully calculated by the English chemist and physicist Henry Cavendish. This constant was later used to calculate the mass of planets. Albert Einstein (U.S.) stated in 1915 that gravitation was a property of space, not a force. (*See also* Planetary Masses)

Additional reading: Asimov. *Asimov's biographical encyclopedia;* p. 148–152; 200–202.

Graviton NUCLEAR PHYSICS

It has been theorized that an accelerated mass or particle emits radiation in a form that has been called gravitational waves. These waves could also have characteristics of particles; such a particle is called a graviton. In 1969 the American physicist Joseph Weber announced that he had detected the effects of gravitational waves. To date, however, no one has been able to duplicate his experiment.

Additional reading: Asimov. *Asimov's new guide to science;* p. 347–348.

Greenhouse Effect METEOROLOGY

Carbon dioxide in the atmosphere has the property of allowing sunlight to pass through it but it absorbs heat radiated from the earth, forming what has been called a "heat trap." This causes the earth's atmosphere to become warmer in the manner of a greenhouse. The Swedish chemist Svante Arrhenius pointed out this process around 1905. In 1932 the American astronomer Walter S. Adams showed that this effect was responsible for the high surface temperature of Venus; the effect was further confirmed in 1962 by the Venus space probe, Mariner II.

Additional reading: Asimov. *Asimov's biographical encyclopedia;* p. 579, 660.

Guanine. *See* **Nucleic Acid.**

Guided Missile. *See* **Missile.**

Gulf Stream. *See* **Oceanography.**

Gunpowder. *See* **Explosive.**

Gymnosperm BOTANY

The name "gymnosperm" was given to conifers and cycads to differentiate them from flowering plants by Scottish botanist Robert Brown during the early 1800s. Brown classified the higher plants in this way with the aid of a **microscope**. (*See also* Angiosperm)

Additional reading: Morton. *History of botanical science;* p. 373–376.

Gyrocompass NAVIGATION

A compass built around a **gyroscope** and which is a true north indicator is called a gyrocompass. The gyroscope is mounted so that it has freedom to move about both a horizontal and a vertical axis. In 1852 the French physicist Leon Foucault showed that a gyroscope could point continuously to true north. In 1903 the German scientist Herman Anschutz-Kaempfer patented the first gyrocompass, and in 1908 Elmer Sperry (U.S.) also patented a gyrocompass. One of his early models was successfully demonstrated on a U.S. Navy vessel in 1911. In the 1920s a mechanical weight was replaced by a fluid device, which in turn has been replaced by electrical sensors and **accelerometers**. (*See also* Inertial Guidance)

Additional reading: *Illustrated science and invention encyclopedia;* v. 9, p. 1159–1160.

Gyroscope ENGINEERING

The gyroscope is a device consisting of a rapidly spinning heavy wheel that maintains its original position when its axis of rotation is moved. This characteristic of being able to resist forces has made the gyroscope valuable for several purposes. When used for stabilizing ships and aircraft, it is known as a **gyrostabilizer.** When used as a compass it is known as a **gyrocompass.** By 1816 an early type of gyroscope was in use, and in 1852 the French physicist Leon Foucault made studies of the gyroscope and gave it its name. (*See also* Aircraft Instrument; Inertial Guidance)

Additional reading: *Illustrated science and invention encyclopedia;* v. 9, p. 1160–1161.

Gyrostabilizer ENGINEERING

Use of gyroscopes to stabilize ships and aircraft has proved very successful. The first application to ships was proposed in 1868 by the English inventor Matthew Watt-Boulton, and in 1903 Otto Schlick (Germany) devised one for a ship. In 1913 Elmer Sperry (U.S.) began installing gyrostabilizers in U. S. warships and merchant vessels. Sperry made use of electric motors to adapt the stabilizer to proper control of the ship. Other versions involve gyroscopically controlled stabilizer fins mounted on a ship's hull and water tanks (called Frahm tank stabilizers) which have gyroscopically controlled air valves. Sperry's first stabilizer for an airplane was installed in 1912; greatly improved models were later developed by him.

Additional reading: *Illustrated science and invention encyclopedia;* v. 9, p. 2182, 2212–2215.

H

Hafnium CHEMISTRY

Hafnium, a strong ductile metal, atomic number 72, is a very difficult element to separate. It is always found in minerals which also contain **zirconium**, primarily zircon and baddeleyite. It was not discovered until 1923 by the Dutch chemist Dirk Coster and the Hungarian chemist Gyorgy Hevesy. Its main use is in nuclear reactors, in vacuum tubes, in light bulbs, and in certain alloys. Chemical symbol is Hf.

Additional reading: *Van Nostrand's scientific encyclopedia;* p. 1444–1445.

Hahnium
CHEMISTRY

Hahnium, atomic number 105, was identified, probably for the first time, in 1970 by a team consisting of A. Ghiorso, M. J. Nurmia, J. A. Harris, K. Eskola, and P. Eskola (all U.S.). It was produced by bombarding **californium** with nitrogen-15 ions. A complicated arrangement was used to capture the atoms sought. They picked the name in honor of noted physicist Otto Hahn, the discoverer of nuclear fission. A conflicting claim of discovery was made by a Soviet team led by G. N. Flerov in 1968. They claim to have produced the element by bombardment of americum by neon ions. The Soviets proposed the name of nielsbohrium, also in honor of a noted physicist. No known use of the element has been reported to date. Chemical symbol is Ha.

Additional reading: *McGraw-Hill encyclopedia of science & technology;* v. 5, p. 2–4.

Hair Dryer. *See* **Dryer.**

Halley's Comet
ASTRONOMY

What we know as Halley's Comet was observed by astronomers dating back to 1532 when the Italian astronomer Girolamo Fracastoro and the German astronomer Peter Apian studied a particular comet. In 1682 the English astronomer Edmund Halley observed a number of comets, then began to study their movements. By 1705 he had noted that the orbits of comets which had appeared in 1456, 1531, and 1607 were similar to the one he had seen in 1682. He reasoned that it must have been the same comet, which was reappearing every 75 or 76 years, meanwhile following an orbit that took it out of sight. Although Halley did not live to see its return in 1758, his prediction was correct and the comet has subsequently followed this schedule, 1986 being its most recent return.

Additional reading: Asimov. *Asimov's new guide to science;* p. 151–153.

Hand Tools
MECHANICAL ENGINEERING

Probably the first hand tool was simply a sharp-edged stick or a piece of stone. Around 20,000 B.C. the bow and arrow were known; the arrows had wooden shafts with a piece of flint fastened to the shaft with resin or pitch. By about 12,000 B.C. pieces of flint were stuck into wood or bone, making a simple saw. Wooden handles came into use around 8000 B.C., as did polished stones. A sickle made of antler bone, having flint teeth, dates back to 5500 B.C. in Palestine. By around 3000 B.C. bronze came into use, but not until 1500 B.C. was it used for tools as well as for ornaments. Bronze axes were used in China around 1100 B.C. By around 800 B.C. iron hand tools had been made by the Assyrians. (*See also* names of specific hand tools)

Additional reading: *Illustrated science and invention encyclopedia;* v. 20, section 1.

Handmill. *See* **Quern.**

Harrow
AGRICULTURE

The triangular or "A" harrow was introduced to American farmers during the 1600s and was used up through the Civil War. It was particularly useful for rocky fields. During the 1840s the two-horse, hinged harrow was used by farmers on well-cleared fields. Its construction was that of a hinged trapezoid with iron or steel spikes. David L. Garver (U.S.) patented a spring tooth harrow in 1869. Finally, American farmers began using disk harrows of Japanese or European origins in the mid-1880s.

Additional reading: Schlebecker. *Whereby we thrive;* p. 104–105, 177–178.

Harvester. *See* **Combine Harvester; Reaper.**

Hay Press
AGRICULTURE

H. L. Emery (U.S.) patented the horse-powered screw-operated hay press in 1853. In 1872 P. K. Dederick (U.S.) introduced the "Perpetual Press," which provided continuous action and thereby sped up pressing of the hay. Steam power was first successfully applied in the United States in 1882. This allowed for more efficient storage and transport of baled hay.

Additional reading: Schlebecker. *Whereby we thrive;* p. 196–198.

Hearing Aid
ACOUSTICS

Miller Reese Hutchison (U.S.) made the first practical electric hearing aid, filing for a patent in 1901. His device was called the "Telephone-Transmitter." In 1920 Earl Charles Hanson (U.S.) invented the first vacuum tube hearing aid. It was commercially distributed in 1921 as the "Vactuphone." Augustus G. Pohlmann (U.S.) and Frederick William Kranz (U.S.) built the first modern bone conduction vibrator in 1923. Finally, in 1953 the first all-transistor hearing aid was introduced in the United States by Microtone.

Additional reading: Berger. *The hearing aid;* p. 52–70, 87, 113–114.

Heart Transplant
MEDICINE

Christiaan Barnard, a South African physician, performed the first successful heart transplant on a human in 1967. His patient, Louis Washkansky, survived for 18 days.

Additional reading: Robinson. *The miracle finders;* p. 203–206.

Heart-Lung Machine
MEDICINE

The first heart-lung machine was invented in 1935 by John H. Gibbon, Jr. (U.S.) and his wife, Mary Gibbon (U.S.). They successfully shut off a cat's heart and lungs, and then kept it alive with their machine. The first time they tried it on a human was in 1952; that patient did not survive. The following year a human patient did live for many years after the procedure, but subsequent patients did not. A simpler, more dependable machine was built in 1955 by American physicians Clarence Walton Lillehei and Richard A. DeWall. Their machine was a bubble-oxygenator that oxygenated the blood and removed the carbon dioxide.

Additional reading: Robinson. *The miracle finders;* p. 170–174.

Heat
PHYSICS

A scientific explanation of heat escaped scientists for many years. At first interest centered on the construction of **thermometers.** By 1760 the Scottish chemist Joseph Black had proved the difference between temperature and heat. Two theories of heat arose; one stated that heat was a fluid named caloric, capable of moving from one object to another. The other theory was that heat was a kind of vibration. In 1798 the English physicist Count Rumford noticed the heat created by boring a cannon and decided that heat was caused by friction. Other researchers in this field were Sir Humphry Davy (England) in 1799 and Frenchmen Jean Fourier in 1822 and Nicolas Carnot in 1824. James Joule (England) showed in 1847 that heat was a form of energy. (*See also* Conservation of Energy)

Additional reading: Asimov. *Asimov's new guide to science;* p. 394–400.

Heat Equivalence. *See* **Conservation of Energy.**

Heat Pump
MECHANICAL ENGINEERING

A heat pump operates between a hot body and a cold body, taking heat from the colder body and pumping it to the hotter body. The same principle as is used by refrigerators or air conditioners is involved; namely, the giving off of heat when a gas condenses. The direction of flow determines whether heat is given off inside or outside the building. The idea is said to have originated with the Irish scientist Lord Kelvin in 1852. It was not until 1927 that a patent for a heat pump was granted to the English inventor T. G. N. Haldane, who used a pump to heat his home and office. In recent years tens of thousands of heat pumps have been installed.

Additional reading: *Illustrated science and invention encyclopedia;* v. 9, p. 1193–1194.

Heating System — MECHANICAL ENGINEERING

Central heating of homes and commercial buildings is a relatively modern practice. Although one example is known from ancient Rome (built around 100 B.C. and known as the hypocaust, in which hot gases from a furnace passed through hollow walls and floors to make them serve as radiant heaters), prior to the nineteenth century most heating was done by fireplaces and stoves. Besides being uncomfortable, it was neither healthful nor efficient. Several types of central heating systems have been in use from around 1780. (*See also* Electric Heating; Heat Pump; Hot Air Heating; Hot Water Heating; Solar Heating; Steam Heating)

Additional reading: *Illustrated science and invention encyclopedia;* v. 9, p. 1241–1242.

Heavy Water. *See* Deuterium.

Helicopter — AEROSPACE ENGINEERING

An important airplane designer was the Russian-born American inventor Igor Sikorsky. As a young boy he tried in vain to make a helicopter with twin rotary blades. He then turned to a successful career designing conventional **aircraft**, notably the famous airline planes of the 1930s known as the Clipper flying boats. Then he turned back to helicopters, building a successful prototype in 1939. Unlike the autogiro, invented in 1928 by the Spanish aeronautical engineer Juan de la Cierva, which could not hover, the helicopter could move vertically and horizontally as well as hover. Sikorsky's was the first successful single-rotor plane; he devoted his post-World War II days to improving his invention.

Additional reading: Feldman. *Scientists & inventors;* p. 278–279.

Helium — CHEMISTRY

Helium, a colorless, tasteless, odorless gas, atomic number 2, is a member of the group known as the noble gases. It was first discovered in 1868 in spectroscopic observations of the sun by Pierre Janssen, a French astronomer, and Joseph Lockyer, an English chemist. In 1895 Sir William Ramsay, a Scottish chemist, made a positive identification while studying a sample of clevite. The only commercial sources are certain natural gas deposits in Utah and Texas. First used for **airships**, its main value now is in aerospace vehicles, in arc welding, and as a refrigerant. Chemical symbol is He.

Additional reading: *Van Nostrand's scientific encyclopedia;* p. 1486–1487.

Hemoglobin — BIOCHEMISTRY

After being one of the first to crystallize it in 1862, German biochemist Ernst Felix Hoppe-Selyer gave the name "haemoglol bin" to the protein responsible for the red color of blood. He is also known for describing the interaction between hemoglobin and oxygen, and set up the first biochemistry laboratory while at the University of Tubingen.

Additional reading: Judson. *The eighth day of creation;* p. 509–511.

Herbarium — BOTANY

Between 1540 and 1542 John Falconer (England) developed the first herbarium on record. His herbarium consisted of a book of dried plants.

Additional reading: Hawks. *Pioneers of plant study;* p. 147.

Herbicide — AGRICULTURE

The first herbicides were developed in the 1940s at Fort Dietrich in the United States. Along with other researchers there, E. J. Kraus (U.S.) studied plant growth hormones. He observed that too much of various hormones could kill plants, and in 1941 suggested that synthetic hormones might make efficient weed killers. In 1941 R. Pokorny (U.S.) synthesized 2,4-dichlorophenoxyacetic acid, which was renamed 2,4-D in 1945. In 1944 Americans John W. Mitchell and Charles L. Hamner noted that 2,4-D was an effective herbicide. 2,4-D was marketed as a herbicide in 1945. (*See also* Pesticide)

Additional reading: Schlebecker. *Whereby we thrive;* p. 270–271.

Heroin — PHARMACOLOGY

In the process of searching for a **morphine** substitute that would be an effective pain reliever yet nonaddictive, Heinrich Dreser (Germany) discovered in 1898 a synthetic derivative of **opium** known as diacetylmorphine. Marketed by F. Bayer and Company as "Heroin," its very powerful addictive qualities were not realized until 1902.

Additional reading: Sneader. *Drug discovery;* p. 72–74.

Hertzsprung-Russell Diagram — ASTRONOMY

The stars have several characteristics by which they can be described. In 1905 the Danish astronomer Ejnar Hertzsprung made some progress on a graph to show stellar properties. In 1913 the American astronomer Henry Russell developed a diagram. The two efforts are merged in what is called the Hertzsprung-Russell Diagram. It compares the absolute brightness of stars with their color (and thus their surface temperature).

Additional reading: Buedeler. *The fascinating universe;* p. 203–205.

Heteromixis — BOTANY

Heteromixis, which involves the fusion of genetically different nuclei, was discovered in fungi by American botanist Albert Francis Blakeslee in 1904. Also known as heterothallism, this process of sexual reproduction was discovered in the Hymenomycetes by M. Bensaude (France) in 1918.

Additional reading: Dunn. *Genetics in the 20th century;* p. 535–536.

High Fidelity Sound. *See* Phonograph.

High Pressure — PHYSICS

The techniques for creating high pressures in laboratories have improved over the years. In the 1880s the French scientist Emile Hilaire Amagat reached 3,000 atmospheres; by 1905 the American physicist Percy Bridgman reached 20,000 atmospheres and eventually 500,000 atmospheres. He changed properties in materials such as ice that remained solid at temperatures above the boiling point of water. The scientist Peter Bell (U.S.) reached 1.5 million atmospheres in the early 1980s.

Additional reading: Asimov. *Asimov's new guide to science;* p. 305, 308.

Hindu-Arabic Numerals. *See* Arabic Numerals.

Histochemistry — BIOCHEMISTRY

Two Frenchmen, J. J. Colin and H. G. de Claubry, discovered the starch-iodine reaction in 1814. When another Frenchman, F. V. Raspail, applied it microscopically, the science of histochemistry, in which the cell's chemical components are studied, was founded.

Additional reading: Bracegirdle. *A history of microtechnique;* p. 95–96.

Histocompatibility Antigen — IMMUNOLOGY

In 1937 P. A. Gorer (England) demonstrated that genetic factors affect the success or failure of tumor transplantation when he discovered the first known histocompatibility antigen in a mouse. He called it "Antigen II."

Additional reading: Dunn. *Genetics in the 20th century;* p. 211, 443.

Histocompatibility Gene
GENETICS

American geneticist George Davis Snell introduced the term "histocompatibility gene" when he located the specific gene sites that affected acceptance or rejection of tissue transplantations in 1948. In 1953 he found that the major histocompatibility complex of the mouse was composed of multiple loci. Snell received a share of the 1980 Nobel Prize in medicine and physiology for his work.

Additional reading: Sturtevant. *A history of genetics;* p. 93–99.

Histology
ANATOMY

French physician Marie Francois Xavier Bichat published his book General Anatomy in 1800 and therein founded the study of histology. Bichat identified 21 types of tissue and showed that organs were composed of different tissue types. Bichat was also the first to use the word "tissue." In 1852 Swiss anatomist Rudolf Albert von Kolliker published the first useful histology textbook.

Additional reading: Magner. *A history of the life sciences;* p. 215–216, 276–278.

Hoe
AGRICULTURE

The earliest hoe appears to have been made of flaked flint in Mesopotamia between 4000 and 3000 B.C. It was used to break up the ground. Flint hoes apparently were also used in Egypt by about 3400 B.C. (*See also* Horse-Drawn Hoe)

Additional reading: Curwen. *Plough and pasture;* p. 29, 64–72.

Holmium
CHEMISTRY

A soft, silver-colored metal, holmium, atomic number 67, was discovered in the mineral erbia in 1878 by the Swiss physicist J. L. Soret and the Swiss chemist Marc Delafontaine. In 1879 Per Cleve, a Swedish chemist, identified it more positively. It is found in such minerals as xenotime, gadolinite, and apatite. Its uses, limited in extent, are in such products as **semiconductors, lasers**, and thermoelectric devices. Chemical symbol is Ho.

Additional reading: *Van Nostrand's scientific encyclopedia;* p. 1509.

Holography
OPTICS

The process of making a three-dimensional reproduction of an image using a **laser** is called holography. The idea originated before lasers were invented, originally with the British physicist Dennis Gabor in 1947, but not until 1965 would lasers be available for its testing, which was then done by the scientist Emmet Leith (U.S.). One half of a laser beam is reflected by a mirror onto a photographic plate while the other half is reflected back by the illuminated objects onto the same plate. This produces an interference pattern known as a hologram. To reconstruct the images, a laser beam is shone onto the plate and the object can be seen. It is 3-dimensional in that observers can see different angles of the object by moving their viewing position.

Additional reading: Asimov. *Asimov's new guide to science;* p. 456.

Homeopathy
MEDICINE

Homeopathy as a field was established in 1810 by Samuel Hahnemann (Germany) when he published *Organon of the Rational Art of Healing.* Hahnemann advocated the administration of small doses of drugs to produce signs and symptoms of a specific disease in healthy patients.

Additional reading: Ackerknect. *A short history of medicine;* p. 145.

Hooke's Law. *See* **Elasticity.**

Hormone
BIOCHEMISTRY

Derived from the Greek word meaning to "rouse to activity," the term "hormone" was introduced in 1905 by the English physiologist Ernest Henry Starling to describe circulating chemical messengers. All hormones together form the body's endocrine system and stimulate metabolic activity.

Additional reading: Crapo. *Hormones: the messengers of life;* p. 11.

Horse Collar
AGRICULTURE

The horse collar was invented in Europe during the Dark Ages in the eighth or ninth century. With its appearance, farmers were able to use horses as draft animals and hitch four or more abreast. This revolutionized the use of animal power because until then farmers could manage only one or two oxen.

Additional reading: Dale. *Topsoil and civilization;* p. 170.

Horse-Drawn Hoe
AGRICULTURE

Jethro Tull (England) invented the horse-drawn hoe in 1699 to assist farmers in destroying weeds and keeping soil between rows in their fields in friable condition. (*See also* Hoe)

Additional reading: Hyams. *Soil and civilization;* p. 263–268.

Hot Air Heating
MECHANICAL ENGINEERING

The hot air heating system consists of a furnace connected by ducts to vents around the structure through which warmer air is blown. This sort of system probably came into use in the United States in the early part of the nineteenth century. The furnace normally burns oil or gas, although until the 1940s coal was commonly used. It has the disadvantage of distributing more dust than other systems.

Additional reading: *Illustrated science and invention encyclopedia;* v. 9, p. 1187–1191.

Hot Water Heating
MECHANICAL ENGINEERING

Hot water heating began in 1777, when a unit was designed for a castle near Paris by the architect G.(?) Bonnemain. The design was similar to that of modern systems. The system did not appear in the United States until around 1840. This type of heating has a central furnace connected by pipes to radiators to which hot water is pumped as needed, returning to the boiler when cool. Furnaces are commonly fired by oil or gas fuels. This system relies on radiant heat as well as induction from the radiators, and it has the advantage of being a clean system.

Additional reading: *Illustrated science and invention encyclopedia;* v. 9, p. 1187–1191.

Hovercraft
AEROSPACE ENGINEERING

Vehicles supported by a cushion of air when in motion, thereby not being in contact with the ground or water, are called hovercraft. The reduced friction allows for efficient use of their power. In modern times one of the earliest models to demonstrate the principles of the vehicle was made by Christopher Cockerell (England) in 1958. The first working hovercraft was tested in 1959, crossing the English Channel in two hours. A model built in the United States the same year traveled at 75 miles per hour. A large British passenger vehicle for use over water is capable of speeds over 90 miles per hour. Hovercraft have found considerable use in military forces as well as in factories and construction projects.

Additional reading: *Illustrated science and invention encyclopedia;* v. 1, p. 45–48, v. 23, p. 196–198.

Human Growth Hormone
BIOCHEMISTRY

Maury Raben (U.S.) developed a method for isolating human growth hormone (HGH) from human cadaver pituitaries in 1956. Soon thereafter John Beck (Canada) showed that HGH promoted growth in human dwarfs.

Additional reading: Crapo. *Hormones: the messengers of life;* p. 122–125.

Hump Yard. *See* **Railroad Switchyard.**

Humulin
PHARMACOLOGY

In 1953 English biochemist Frederick Sanger determined the exact order of amino acid chains for the whole **insulin** molecule. Sanger, awarded the 1958 Nobel Prize for medicine and physiology, had determined the sequence of a protein for the first time. His work stimulated the Eli Lilly International Corporation in the United States to use a genetic engineering technique known as recombinant DNA in the production of human insulin. In 1982 the pharmaceutical company marketed its product under the trade name "Humulin."

Additional reading: Sneader. *Drug discovery;* p. 216–217.

Hybridoma. *See* Monoclonal Antibody.

Hydrochloric Acid
CHEMISTRY

An extremely corrosive substance, hydrochloric acid, also known as muriatic acid, was prepared as early as 1611 by Andreas Libau (Germany). In 1775 the English chemist Joseph Priestly discovered the gaseous form of the compound. However, it was not until 1810 that the English chemist Sir Humphry Davy proved that oxygen was not a constituent of hydrochloric acid, despite the opinion of many chemists that all acids contained oxygen. Hydrochloric acid is commonly manufactured by electrolysis of brine (salt water), and it has many industrial uses.

Additional reading: *Van Nostrand's scientific encyclopedia;* p. 1542–1543.

Hydrodynamics. *See* Fluid Mechanics.

Hydroelectric Station
ELECTRICAL ENGINEERING

Turbines driven by water power to create electricity, known as hydroelectric generators, originated in France in 1869 when the paper manufacturer Aristide Berges used a cascade of water in the Alps to drive electric generators. He used a waterfall about 600 feet high to generate 1.5 kilowatts of energy. Current hydroelectric plants include the "pumped storage" concept in which water is pumped to a reservoir at a higher elevation during times of abundant electric power. When peak demand for power is reached, the water is released to drive a hydroelectric generator and supply extra power.

Additional reading: *Illustrated science and invention encyclopedia;* v. 18, p. 2499–2502.

Hydrofoil
NAVAL ARCHITECTURE

The hydrofoil is a ship which has fins or foils attached to the hull so that at a certain speed the hull is lifted almost completely out of the water due to hydrodynamic pressure. There are two types of foils—those that are fully submerged in water all the time and the surface-piercing system in which the foils rise and fall depending upon wave heights. The latter system is best used in quiet coastal waters. The principles were established by Leon Farcot (France) in 1869. One of the early inventors was Alexander Graham Bell (U.S.), who with his assistant Casey Baldwin (U.S.) produced a model in 1918 that reached a speed of over 70 miles per hour. In 1962 a Boeing Aerospace Company model hydrofoil submarine chaser was built using electronic control of foil surface angles.

Additional reading: *Illustrated science and invention encyclopedia;* v. 3, p. 268; v. 23, p. 199.

Hydrogen
CHEMISTRY

The discovery of a gas, later to be named hydrogen, atomic number 1, was made by the English chemist Henry Cavendish in 1766, although it was not named until nearly twenty years later. It is a tasteless, colorless, odorless gas, and is the lightest element, weighing only about one fourteenth as much as an equal volume of air. It is produced commercially as a by-product of the manufacture of sodium hydroxide as well as by passing steam over hot coke or iron. It is used in many industrial compounds as well as in welding.

Additional reading: Downs. *Landmarks in science;* chap. 41.

Hydrogen Bomb. *See* Thermonuclear Weapon.

Hypernucleus. *See* Lambda Particle.

Hypnosis
PSYCHOLOGY

Hypnosis was introduced in 1774 by Franz Anton Mesmer (Germany), who treated psychosomatic illnesses by suggestion, and believed the process involved animal magnetism. In 1841 James Braid, a Scottish surgeon, coined the term "hypnotism" from the Greek word for sleep. Braid observed that the process involved a suspension of the conscious mind. Jean Martin Charcot, a French physician, began using Braid's techniques in 1872, and in 1885 introduced them to an Austrian psychiatrist named Sigmund Freud. This exposure induced Freud to examine the psychological aspects of mental disorders. By the 1890s Freud had abandoned hypnotism as a technique. Between 1880–1882 Josef Breuer (Austria) introduced the use of hypnosis as a treatment for hysteria. (*See also* Catharsis)

Additional reading: Murray. *A history of Western psychology;* p. 293–297.

Hypocaust. *See* Heating System.

I

ICBM. *See* Missile.

Ice Age. *See* Glacier.

Ice Island
OCEANOGRAPHY

Research stations located on floating ice in the Arctic are known as drifting ice stations. They provide a place for making meteorological and oceanographic observations in an area inaccessible to surface ships. Their predecessors were strengthened ships purposely frozen into the ice so as to drift with it, the early example being the Norwegian North Polar Expedition led by Fridtjot Nansen in 1893–1896 and in 1918–1924 by Roald Amundsen. The first drifting ice island was established by the U.S.S.R. in 1936, where a party of four led by Ivan Pabanin stayed for nine months after having been brought by an airplane landing on the ice. In 1952 the United States established Fletcher's Ice Island. It was operated for several years; others have since been established.

Additional reading: Fairbridge. *Encyclopedia of oceanography;* p. 232–233.

Iconoscope. *See* Television.

Immunization
IMMUNOLOGY

In 1798 Edward Jenner, an English physician, published "An Inquiry into the Causes and Effects of the Variolae Vaccinae" and therein described his successful experiments with vaccinations. As a result of his research, Jenner initiated the science of immunology by establishing the principle of active immunization. (*See also* names of specific vaccines)

Additional reading: Asimov. *Asimov's new guide to science;* p. 677–688.

Immunological Tolerance
IMMUNOLOGY

Australian physician Sir Frank Macfarlane Burnet suggested in 1949 that human capacity to produce **antibodies** against the **proteins** of another human might not be inborn but instead might be acquired during fetal development. Following up on Burnet's suggestion, English biologist Sir Peter Brian Medawar inoculated mice embryos with tissue cells from another strain of mice. He confirmed Burnet by finding that antibodies had not yet been formed. Medawar reported his research in 1953. The two men shared the 1960 Nobel Prize in medicine and physiology for their discovery of acquired immunological tolerance. (*See also* Antibody)

Additional reading: Sourkes. *Nobel prize winners in medicine and physiology 1901-1965;* p. 379–389.

In-flight Refueling
AEROSPACE ENGINEERING

The first use of a pipe connecting two airplanes in a refueling operation took place in 1923 when two United States Army officers accomplished it—the participants were Captain Lowell Smith and Lieutenant J. P. Richter. Since then two systems, the probe and drogue method and the flying boom method, have been developed for the refueling of single-occupant planes.

Additional reading: *Illustrated science and invention encyclopedia;* v. 15, p. 1960–1962.

Incandescent Lamp. *See* Electric Light.

Incinerator. *See* Waste and Sewage Disposal.

Inclined Plane. *See* Gear.

Incubator
MEDICINE

In 1884 an incubator warmed by kerosene lamps appeared in Paris at La Maternite. American physician Julius H. Hess designed an electric incubator for premature infants and filed for a patent in 1933.

Additional reading: Davis. *Medicine and its technology;* p. 50.

Indium
CHEMISTRY

Indium, atomic number 49, is a silver-white metal which is softer than lead, malleable and ductile. It was discovered in 1863 during a spectographic study of zinc ores by the German physicist Ferdinand Reich and the German chemist Theodor Richter. The main source is through treatment of the flue dust in zinc production. Its chief use is in semiconductors, in certain alloys, and in nuclear reactors. Chemical symbol is In.

Additional reading: *Van Nostrand's scientific encyclopedia;* p. 1591–1592.

Induction. *See* Automobile Ignition System; Electromagnetism.

Inertial Guidance
NAVIGATION

Inertial guidance platforms involve the use of gyroscopes and accelerometers to serve as guidance systems for missiles, space vehicles, submarines, and other vehicles. They have proved to be very accurate. The first application was in 1944 during World War II in German V2 missiles. By 1948 the United States had developed full inertial systems. In 1958 such a system was proved accurate in navigating a U. S. submarine under the Arctic ice cap. Inertial guidance serves as a navigational device without reference to external factors.

Additional reading: Illustrated science and invention encyclopedia; v. 10, p.1258-1261.

Infrared Photography
GRAPHIC ARTS

Infrared photography became possible in 1942 when an American firm developed film for taking color pictures sensitive to this type of radiation. Now a common method, infrared photographs are particularly good for penetrating haze and for such diverse uses as showing the presence of oil slicks, the status of farm crops, and the severity of storms.

Additional reading: *Illustrated science and invention encyclopedia;* v. 1, p. 22–27; v. 23, p. 179–180.

Infrared Radiation
PHYSICS

Through study of the spectrum produced when sunlight passed through a prism, German-born Sir William Herschel (England) discovered around 1800 that when he measured the temperature of the different colors the highest temperatures were found beyond the red end of the spectrum. He realized radiation invisible to the eyes was being transmitted from the sun. This radiation was named infrared radiation.

Additional reading: Asimov. *Asimov's new guide to science;* p. 60–61.

Insecticide. *See* Pesticide.

Instant Camera
GRAPHIC ARTS

A means of providing literally instant photographs was provided by the work of the American physicist Edwin Land who developed what was clearly a popular invention, the instant camera, in 1947. Essentially, the camera originally used a double roll of film, one part being ordinary negative film and the other a positive paper; between the two were chemicals which developed the image and formed the print. In 1963 color film was introduced. Improvements in 1972 eliminated the need to peel the two layers apart. Then in 1977 a system for processing motion pictures allowed the film to be exposed, processed, and viewed without being taken from a cassette.

Additional reading: Feldman. *Scientists & inventors;* p. 302–303.

Insulin
BIOCHEMISTRY

Romanian physiologist Nicolas Paulesco discovered insulin in the human pancreas in 1921. Within the same year Sir Frederick Grant Banting (Canada), Charles H. Best (Canada), John J. MacLeod (Scotland), and James Collip (Canada) developed a method of extracting it from the human pancreas and then purifying it so it could be injected into the blood of diabetics. They demonstrated its safety by injecting it into Banting and Best. By 1922 Banting's group began the first insulin therapy on a young Canadian diabetic, Leonard Thompson. John Jacob Abel, an American biochemist, prepared it in crystalline form for the first time in 1925. Banting and MacLeod won the 1923 Nobel Prize in medicine and physiology for their work. (*See also* Humulin)

Additional reading: Crapo. *Hormones: the messengers of life;* p. 129–133.

Integrated Circuit
ELECTRONICS

Efforts to miniaturize electronic circuits in the last 40 years were greatly aided by the invention in 1948 of the **transistor,** allowing virtual elimination of bulky electron tubes. Ways to reduce the size of other components led to miniature circuits. The idea for integrated circuits came from the English radar engineer W. A. Dummer in 1952. By 1960 improved methods of making circuit components on silicon chips through precise photoetching techniques brought even greater size reduction; these were called integrated circuits. In 1960 a miniature circuit might have had one transistor in a 5-millimeter cube; by 1964 an integrated circuit could contain 100 transistors in that space and by 1975 10,000. The creation of the **microchip** has resulted from this method.

Additional reading: Clark. *Works of man;* p. 287–288.

Intelligence Quotient
PSYCHOLOGY

The concept of "intelligence quotient" was introduced in 1912 by William Stern (Germany). He suggested that the mental age obtained from an **intelligence test** be divided by chronological age to yield IQ. This figure would express the individual's intelligence in relation to others of the same biological age.

Additional reading: "20 discoveries that changed our lives"; p. 55–57.

Intelligence Test
PSYCHOLOGY

French psychologists Alfred Binet and Theophile Simon published the first battery of tests written to measure intelligence in 1905. In 1916 Lewis M. Terman (U.S.) translated a 1908 edition of Binet's test, and established norms for American children. Terman then published what was known as the Stanford-Binet Intelligence Scale. (*See also* Mental Test)

Additional reading: "20 discoveries that changed our lives"; p. 55–57.

Intercontinental Ballistic Missile. *See* Missile.

Interferometer PHYSICS

A means for accurately measuring the wave length of light was provided in 1893 when the American physicist Albert A. Michelson invented the interferometer. Modern types now operate in other ranges, such as acoustical and radio interferometers.

Additional reading: Asimov. *Asimov's new guide to science;* p. 382–383.

Interferon BIOCHEMISTRY

In 1957 Alick Isaacs (England) and Jean Lindernmann (England) showed that cells released a **protein** called "interferon" that was capable of inhibiting invading viruses. Interferon was also shown to multiply much faster than antibodies. In 1977 Sydney Pestka (U.S.) purified interferon and established that it existed as multiple, closely related proteins.

Additional reading: Asimov. *Asimov's new guide to science;* p. 687–688.

Internal Combustion Engine MECHANICAL ENGINEERING

The success of the automobile depended upon the invention of the internal combustion engine (one in which the fuel is burned internally, as in a gasoline engine). In 1849 Stuart Perry (U.S.) patented an internal combustion engine, that ran on turpentine. The French engineer Etienne Lenoir designed a successful engine about 1860, although it was bulky and inefficient, running on a mixture of air and coal gas. He did some work toward putting it in an automobile in 1862. A major improvement was the four-stroke internal combustion engine of the German engineer Nikolaus Otto in 1876. A major change in design of the four-stroke model came in 1957, when the German inventer Felix Wankel produced a rotary-piston engine that contained a triangular-shaped piston rotated within a casing.

Additional reading: Feldman. *Scientists & inventors;* 164–165, 180–181, 294–295.

Internal Combustion Tractor AGRICULTURE

In 1889 a tractor with an internal combustion engine was built in the United States, but it had too little power to run a plow. John Froelich (U.S.) introduced the first useful gasoline tractor in 1892. This began the era of agricultural industrialization when other sources of power were used in place of animals. In 1901 Americans Charles Hart and Charles Parr built the first successful Hart-Parr tractor. They opened the first tractor company in Charles City, Iowa, in 1903 and gave the name "tractor" to their machine in 1906. (*See also* Track Tractor)

Additional reading: Schlebecker. *Whereby we thrive;* p. 199–205.

Internal Secretion ENDOCRINOLOGY

Claude Bernard (France) first introduced the term "internal secretion" when describing liver physiology in 1855. In 1895 the English physiologist Sir Edward Albert Sharpey-Schafer promoted the idea that these secretions come from glands upon his discovery that dogs develop diabetes when their pancreas glands are removed. He also found that thyroid glands produce internal secretions in humans. The development of this theory eventually led to important advances in the understanding of **hormone** physiology.

Additional reading: Crapo. *Hormones: the messengers of life;* p. 7.

Intrauterine Device. *See* **IUD.**

Introvert. *See* **Extrovert-Introvert.**

Invertebrate Biology BIOLOGY

Between 1815 and 1822 French naturalist Jean Baptiste Lamarck published his seven-volume work entitled *Natural History of Invertebrates,* and began modern invertebrate biology. Lamarck was the first person to use the terms "vertebrate" and "invertebrate."

Additional reading: Wightman. *The growth of scientific ideas;* p. 402.

Iodine CHEMISTRY

Iodine is a bluish-black solid, atomic number 53, having a metallic luster and some other properties of a metal. It is in the halogen class of elements and was discovered in 1811 by the French chemist Bernard Courtois. Iodine is widely distributed in nature, including rocks, soils, and sea water. It is widely used in medicine, in industrial compounds, and in foods. Chemical symbol is I.

Additional reading: *Van Nostrand's scientific encyclopedia;* p. 1651–1652.

Ion CHEMISTRY

A study of the behavior of the passage of electricity through solutions in certain cases led a Swedish chemist, Svante Arrhenius, to propose that molecules of some compounds break up into charged particles when put in a liquid. When first proposed in 1884, this idea was considered too radical. His concept of ions being atoms bearing electric charges was generally ridiculed until the discovery of the **electron** showed him to be right. Later study showed that high temperatures or certain radiation can also ionize neutral atoms. In 1922 ions were discovered in the atmosphere in what is now called the **ionosphere.**

Additional reading: Asimov. *Asimov's new guide to science;* p. 220–221.

Ionosphere GEOPHYSICS

Curiosity about the reason radio reception faded out at night caused the English physicist Sir Edward Appleton to study the phenomenon. In 1922 he decided it was caused by the existence of two different routes for the radio waves to follow, involving a direct path for some and reflection from the upper atmosphere for others. In 1924 he located and measured the reflecting layer, the **Kennelly-Heaviside layer,** about 65 miles high. He named that part of the atmosphere the ionosphere because of the high number of ions present there. (This term had been proposed by the Scottish physicist Robert Watson-Watt.) Different levels within the layer are today designated by letters. Layers some 140 miles high have been named the Appleton layers.

Additional reading: Asimov. *Asimov's new guide to science;* p. 220–222.

Iproniazid PHARMACOLOGY

Iproniazid was developed in the 1950s by the Hoffmann-LaRoche Pharmaceutical Company for treating tuberculosis. It is now known as the first of the antidepressive drugs, however, due to the work in 1956 of an American psychiatrist, Nathan S. Kline. Kline had found evidence of an antidepressive action in the earliest literature describing Iproniazid. He began testing its effectiveness with depressive patients and, in 1957, was able to report great success in treating them with it.

Additional reading: Robinson. *The miracle finders;* p. 228–235.

Iridectomy. *See* **Ophthalmology.**

Iridium CHEMISTRY

Iridium, atomic number 77, is a member of the platinum group of metals. It is the most dense of all elements, has the greatest resistance to corrosion, and has a great mechanical strength at high temperatures. It was discovered in 1803 by Smithson Tennant, an English chemist. It is used chiefly in alloys or (in pure metallic form) in small crucibles for high temperature. Chemical symbol is Ir.

Additional reading: *Van Nostrand's scientific encyclopedia;* p. 1659.

Iron CHEMISTRY

One of the world's most important metals, iron, atomic number 26, is a silver-white metal that is both ductile and malleable. It is likely that it has been in use since about 6000 B.C., and was probably discovered by accident. Iron is the fourth most common element in the earth's crust, being found all over the world. Most of the iron made today is used in the production of steel. Chemical symbol is Fe. (*See also* Steel)

Additional reading: *Van Nostrand's scientific encyclopedia;* p. 1660–1664.

Iron Lung
MEDICINE

J. M. Howe (U.S.) invented a device during the 1840s containing hot water from which patients could inhale warm air. Eventually these portable machines were replaced by large, stationary units into which patients were placed to aid them when suffering from lung diseases. During the 1860s, a forerunner of the iron lung, the spirophore, was used to revive asphyxiated patients and newborn babies. In 1894 American industrial hygienist Philip Drinker invented the iron lung to act as an artificial respirator.

Additional reading: Davis. *Medicine and its technology;* p. 196–201.

Iron Smelting
METALLURGY

The process of heating iron to produce the liquid metal was for many years dependent upon charcoal to fire the furnaces. In 1709 the English iron manufacturer Abraham Darby made a major change by switching to coke (produced from coal) as his fuel. It not only stopped the depletion of British forests (as a raw product for charcoal) but it also resulted in higher quality iron. Several techniques have been developed over the years for the treatment of iron in the making of **steel.**

Additional reading: Feldman. *Scientists & inventors;* p. 50–51.

Irrigation. *See* Canal.

Islets of Langerhans
ENDOCRINOLOGY

While studying the microscopic structure of the pancreas in 1869, Paul Langerhans, a German medical student, discovered cells scattered throughout the organ that differed from all the others. In 1893, Gustave-Edouard Laguesse (France) suggested that these cells, known as the Islets of Langerhans, secreted something that prevented diabetes. That secretion turned out to be **insulin.**

Additional reading: Crapo. *Hormones: the messengers of life;* p. 130–131.

Isomer
CHEMISTRY

The fact that two or more compounds can have the same chemical formula yet have different structures and properties was first discovered in 1824 by the French chemist Joseph Gay-Lussac, when reading papers for a journal he edited. He noticed that a paper on fulminates used the same formula as a different paper on cyanates. At first some scientists like the influential Swedish chemist Jons Berzelius found this hard to accept. Finally, in 1830, Berzelius discovered it for himself. He gave such compounds the name isomers (from the Greek words meaning "equal parts"). The greater the number of atoms in a molecule, the greater the number of isomers that could exist.

Additional reading: Asimov. *Asimov's new guide to science;* p. 510, 512.

Isoniazid
PHARMACOLOGY

In 1912 two Czechoslovakian chemists, Hans Meyer and Josef Mally, synthesized a new chemical called isonicotinic acid hydrazide. That chemical is now known as "isoniazid." Isoniazid was first tried as a cure for patients suffering from tuberculosis in 1951 at Sea View Hospital in New York. Irving J. Selikoff (U.S.) and Edward H. Robitzek (U.S.) ran those tests on people who were not responding to treatment with **streptomycin.**

Additional reading: Robinson. *The miracle finders;* p. 9–13.

Isotope
ATOMIC PHYSICS

Studies of atomic weights of the elements occasionally showed small but definite variations. In 1913 the English chemist Frederick Soddy proposed that there were different atoms of elements which vary in atomic weight but not in chemical properties; he called them isotopes. His work was verified by the American chemist Theodore Williams later in 1913. In time atomic masses were determined by electromagnetic methods. (*See also* Radioisotope)

Additional reading: Asimov. *Asimov's biographical encyclopedia;* p. 618–619, 664–665.

IUD
MEDICINE

The first IUD or intrauterine device designed to block **fertilization** of the human egg by the **sperm** was built by R. Richter, a German physician, in 1909. Richter's IUD was constructed of silkworm gut. In 1923 Karl Pust (Germany) developed the first metallic IUD, built of silver wire.

Additional reading: Robinson. *The miracle finders;* p. 273–274.

J

Jacquard Loom
TEXTILES

Efforts to simplify the weaving of cloth having patterns or designs date back to the Middle Ages. At first this involved the use of an assistant to pull the cords by hand to match a given pattern or design, but in 1725 the French inventor Basile Bouchon devised a loom in which a roll of paper punched with holes would produce a pattern, eliminating mistakes but still requiring an assistant. In 1801 Joseph Jacquard (France) was called in to improve a loom made in 1745 by Jacques de Vaucanson (France) which eliminated the assistant but was too complicated to be practical. Jacquard's model was simple and became an instant success—by 1812 there were 11,000 in France alone. It is still in use today with minor changes. The loom was controlled by stiff **punched cards.**

Additional reading: *Illustrated science and invention encyclopedia;* v. 10, p. 1295–1297.

JATO. *See* Jet Takeoffs

Jet Engine
AEROSPACE ENGINEERING

Research on a jet engine began in 1921 when Sir Frank Whittle, an English engineer, began tests on a turboprop engine, one in which a rotor, powered by hot gases striking a series of blades, drove a propeller. These engines later lost out to pure jet engines. In 1930 Whittle patented a jet engine in which air was compressed, then routed into a combustion chamber to be ignited. The hot gases would leave through the exhaust with great force, the reaction to which would push the plane forward. He tested his first engine in 1937 but not until 1941 was a flight made. In the meantime the German engineer Hans van Ohain had developed a jet engine which led to the first jet-powered flight in 1939. A ramjet engine is used in guided missiles, as is a similar type known as a polarjet.

Additional reading: Feldman. *Scientists & inventors;* p. 300–301.

Jet Stream
METEOROLOGY

During World War II high level bombers discovered very strong winds up to 500 miles per hour. Now called jet streams, one is found in each hemisphere. They were studied by the Swedish-born meteorologist Carl-Gustaf Rossby (U.S.) in the late 1940s.

Additional reading: Asimov. *Asimov's new guide to science;* p. 212.

Jet Takeoffs
AEROSPACE ENGINEERING

During World War II both Germany and the United States developed systems which would give heavy aircraft extra power to take off from short runways. Small rockets attached to the plane were set off electrically. Some were permanently fastened, and others were dropped by parachute. Around 1942 the American inventor Robert Goddard worked on such devices for use on aircraft carrier planes. These inventions, called jet-assisted takeoff (or JATO) devices, are sometimes referred to as rocket-assisted takeoff devices (or RATO).

Additional reading: *Illustrated science and invention encyclopedia;* v. 10, p.1299.

Jupiter ASTRONOMY

Jupiter is the most massive of all the planets and by virtue of that size appears very bright in the sky despite never being closer to the earth than 390 million miles. The first to study it with a telescope was the Italian astronomer Galileo Galilei in 1610; he noted four satellites revolving around it and calculated their motions. In 1668 Giovanni Cassine (France) established Jupiter's period of rotation and also a table of motion of its satellites, which later helped in calculation of the speed of light. In 1733 the English astronomer James Bradley measured Jupiter's diameter. A fifth satellite was discovered in 1892 by the American astronomer Edward Barnard. Since then eleven more satellites were discovered. Jupiter probes in the 1970s added much to our knowledge of it.

Additional reading: Asimov. *Asimov's new guide to science;* p. 127–134.

K

Kayon NUCLEAR PHYSICS

Another particle that belongs to the meson group is the K-meson or Kayon. It was first detected in 1952 by two Polish physicists, Marran Danysz and Jerzy Pniewski. Kayons have a mass about 970 times that of an electron and last for about a millionth of a second.

Additional reading: Asimov. *Asimov's new guide to science;* p. 359.

Kelvin Temperature Scale PHYSICS

The law of gases discovered in 1787 by the French physicist Jacques Charles was puzzling to the Scottish scientist Lord Kelvin. In 1848 Kelvin realized that at -273 degrees centigrade the energy of the gas molecules would be zero. He theorized that this would hold for any substance and that this temperature represented the lowest temperature possible, or absolute zero. He devised a new temperature scale with degrees the same size as those of the centigrade plus 273. The letter K is used to identify the scale, called "degrees Kelvin."

Additional reading: Feldman. *Scientists & inventors;* p. 170–171.

Kennelly-Heaviside Layer GEOPHYSICS

The existence of an electronically charged layer in the upper atmosphere was predicted independently in 1902 by two electrical engineers, Oliver Heaviside (England) and Arthur Kennelly (U.S.). Heaviside had made his predictions on the basis of mathematical analysis, whereas Kennelly based his on analyzing what could have caused radio signals sent by Guglielmo Marconi (Italy) to reach all the way to Newfoundland from England in 1901. Kennelly theorized the waves had to have been reflected from the upper atmosphere. This was shown to be a fact when studies of the **ionosphere** in 1922 located the layer some 65 miles high.

Additional reading: Asimov. *Asimov's new guide to science;* p. 221–222.

Kidney Transplant MEDICINE

The first kidney transplant occurred in 1902 when Viennese surgeon E. Ullman transplanted a kidney from one dog to another. In 1936 Soviet surgeon U. Voronoy transplanted a kidney from a person soon after death to someone suffering from mercury poisoning. The patient lived for two days. The first successful kidney transplant between two living humans was performed in 1950 by American physician Richard H. Lawler. In 1963 American physician Thomas E. Starzl successfully tackled the problem of organ rejection, the main difficulty with kidney transplants, by utilizing the immunosuppressant drugs prednisone and azathioprine. The latter drug was discovered by George Hitchings (U.S.) in the 1950s.

Additional reading: Robinson. *The miracle finders;* p. 74–81.

Kiln MECHANICAL ENGINEERING

The early use of pottery required a means for drying and hardening the clay. As far back as 3000 B.C. kilns were used in Mesopotamia for pottery making. Early kilns were unable to reach temperatures to vitrify the clay so that it achieved a glossy quality. Greek decorations in the fifth and sixth centuries B.C. involved successful firing of the surfaces so as to permit elaborate decorations. Brick kilns dated from 60 A.D. in Rome, continuing throughout Europe. By about 1400 in England three to four laborers could produce 100,000 bricks per year using huge rectangular kilns. In the 1800s coal-fired bottle-shaped kilns were used. In 1858 the tunnel kiln was built by the German railway engineer A. Hoffmann, providing a continuous process.

Additional reading: Derry. *Short history of technology;* p. 18, 75–77, 87–88, 94, 590–591.

Kinase BIOCHEMISTRY

German-born biochemist Otto Fritz Meyerhof introduced the term "kinase" in 1927 in his description of the processes enabling skeletal muscles to form lactic acid from free hexoses. Kinase referred to the group of enzymes that he found catalyzed phosphorylation reactions.

Additional reading: Kalckar. "The discovery of hexokinase"; p. 291–293.

Kinescope. *See* **Television.**

Kinetic Theory of Gases PHYSICAL CHEMISTRY

Closely related to various **gas laws** was research on **diffusion** of gases. Then around 1860 the Scottish mathematician James Clerk Maxwell used his mathematical skill to predict that molecules of gases in a closed container moved in all directions and at all velocities, rebounding from the container walls with perfect elasticity. Working independently of Maxwell, the Austrian physicist Ludwig Boltzmann came to the same conclusions, and the two share the credit for this discovery. They thus showed heat to be a function of the motion of molecules. (*See also* Diffusion)

Additional reading: Asimov. *Asimov's biographical encyclopedia;* p. 359, 454–455, 500.

Knowledge-Based System. *See* **Expert System.**

Krebs Cycle. *See* **Citric Acid Cycle.**

Krypton CHEMISTRY

Krypton, a colorless, odorless gas, atomic number 36, is a member of the noble gas class. It was discovered in 1898 by the Scottish chemist Sir William Ramsay and the English chemist Morris Travers. The major source is the liquefaction and distillation of air. One of its major uses is with other gases in fluorescent lights. Chemical symbol is Kr.

Additional reading: *Van Nostrand's scientific encyclopedia;* p. 1721.

Kymograph MEDICINE

Around 1705 Stephen Hales, an English botanist and chemist, was the first to measure the blood pressure of various animals. Hales designed and attached the first **manometer**, in the form of a glass tube, to a vein or artery of a living animal such as a horse, and then recorded the height to which its blood rose. In 1828 Jean Leonard Marie Poiseuille, a French physician, developed a mercury manometer to improve on Hales' instrument. Karl Friedrich Wilhelm Ludwig, a German physiologist, built a kymographion in 1847 which allowed blood pressure value to be recorded on a rotating drum. Ludwig's kymograph demonstrated that blood circulation could be explained by mechanical forces. (*See also* Sphygmomanometer)

Additional reading: Woglom. *Discoverers for medicine;* p. 11–29.

L

L-D Process. *See* **Basic Oxygen Process.**

L-Dopa
PHARMACOLOGY

During the 1960s, Greek-born physician George C. Cotzias (U.S.) discovered how to get past the blood-brain barrier with L-dopa (also known as Levodopa) to treat patients suffering from Parkinson's disease, which attacks nerve cells in the brain. Cotzias began to treat patients in 1966, and made his research public in 1969.

Additional reading: Robinson. *The miracle finders;* p. 29–34.

Lambda Particle
NUCLEAR PHYSICS

When a lambda particle, a member of the class of subatomic particles called baryons, replaces a **neutron** in a nucleus, it forms a hypernucleus, which exists for less than a billionth of a second. The hypertritium nucleus was the first found, discovered in 1952 by the Polish physicists Marren Danysz and Jerzy Pniewski.

Additional reading: Asimov. *Asimov's new guide to science;* p. 360–361.

Lamp. *See* **Electric Light; Oil Lamp.**

Lanthanum
CHEMISTRY

Lanthanum, atomic number 57, is a soft, silver-gray metal, which must be handled in an inert atmosphere. It was first identified in 1839 by the Swedish chemist C. G. Mosander. The main commercial source is extraction from bastnasite. It is used in alloys, both those involved in making permanent magnets, as well as those used at high temperatures. Chemical symbol is La.

Additional Reading: *Van Nostrand's scientific encyclopedia;* p. 1727–1728.

Laryngoscope
MEDICINE

In 1854 Spanish-born singing teacher Manuel Patricio Rodriguez Garcia (England) designed the first laryngoscope that allowed a clear view of the glottis at work. Garcia demonstrated that the vocal cords were responsible for the voice and its register, falsetto, tones, and the like. With Garcia's laryngoscope, it became possible to see any obstructions occurring in the larynx.

Additional reading: Woglom. *Discoverers for medicine;* p. 84–99.

Laser
OPTICS

The laser, a device for producing an intense parallel beam of light at a practically pure wavelength, was invented by Theodore H. Maiman (U.S.) in 1960, based on work done on the **maser** as well as the studies made by Charles Hard Townes and A. L. Schawlow (both U.S.) in 1958. "Laser" is an acronym for "light amplification by stimulated emission of radiation." Lasers may use solids, liquids, or gases. The parallel nature of their light is shown by the experiment in 1962 when a laser beam was sent to the moon, covering an area only two miles wide. Lasers have many applications for both civilian and military purposes. One of the most publicized is the controversial Strategic Defense Initiative or "Star Wars" plan for defense against missiles. (*See also* Holography)

Additional reading: Asimov. *Asimov's new guide to science;* p. 453–456.

Laser Disk
ELECTRONICS

Laser disks, often called compact disks (CDs) or video disks, use a laser to read information recorded on flat disks, the data taking the form of microscopically small pits in the surface. There is no surface wear and high quality transmission is possible. They may be used to store graphic symbols and sound or alphanumeric characters for computers. They were developed around 1972 by the Philips Company (Holland). A competing non-laser compact system was developed at about the same time by the RCA Company (U.S.). It used a stylus to read grooves, much like a conventional phonograph player.

Additional reading: *Illustrated science and invention encyclopedia;* v. 29, p. 2563–2564; v. 22, p. 96–98.

Latitude and Longitude
GEODESY

The method of locating an object according to its latitude and longitude was probably first developed by the Greek astronomer Hipparchus around 134 B.C. His work was improved by Ptolemy, a Greek geographer and astronomer, who adapted the system to measurements on the earth sometime around 150 A.D.

Additional reading: Downs. *Landmarks in science;* chap. 10.

Laudanum. *See* **Opium.**

Laughing Gas. *See* **Nitrous Oxide.**

Lawnmower
MECHANICAL ENGINEERING

For centuries lawns were cut by scythes, then a mechanized mower was patented in 1830 by the English inventor Edwin Budding. His machine used helical cutting edges. Rotary mowers used a single rotating blade and were power driven. Gas engines were used around 1919, particularly in the United States. In 1958 the electric lawnmower was devised by an inventor named Wolf. An air cushion mower has been introduced in England in recent years—it floats about quarter of an inch above the ground.

Additional reading: *Illustrated science and invention encyclopedia;* v. 10, p. 1239–1241.

Lawrencium
CHEMISTRY

A radioactive metal of the actinide series, lawrencium, atomic number 103, was discovered in 1961 by the American chemists A. Ghiorso, T. Sikkeland, A. Larsh, and R. Latimer. It is produced by bombardment of appropriate elements. No use has been reported for it to date. Chemical symbol is Lw.

Additional reading: *Van Nostrand's scientific encyclopedia;* p. 1743.

Lead
CHEMISTRY

Lead, atomic number 82, is one of the oldest metals known to man, having been used at least as early as 3500 B.C. by the Egyptians. It is a very soft material and has a very low melting point. Lead is principally found in the mineral galena. It is one of the four most largely produced and used metals; its chief applications are in construction work, in alloys, and in storage battery plates, and as a protective coating for iron and steel. Chemical symbol is Pb.

Additional reading: *Van Nostrand's scientific encyclopedia;* p. 1743–1746.

Leaf Knife
AGRICULTURE

The leaf knife was invented in Europe between 500 and 400 B.C. It was used as a tool to strip leaves from tree branches to provide fodder during the winter months when cattle had to be housed indoors. The leaf knife consisted of a **sickle** blade set in a line with the handle instead of at an angle with it.

Additional reading: Curwen. *Plough and pasture;* p. 97, 116–121.

LED. *See* **Light Emitting Diode.**

Lens
OPTICS

The development of optical glass came many centuries after the first production of glass for nonscientific uses. As early as 1000 A.D. the Islamic philosopher Ibn-al-Haitham knew about the properties of lenses. It was not until about 1280 that the first spectacles were invented; the art was well established at Venice by the early fourteenth century. Galileo Galilei (Italy) made a working **telescope** by 1609, and by 1637 Rene Descartes (France) showed how to correct for spherical aberration. In 1671 Sir Isaac Newton discovered the causes of chromatic aberration, although it was not until 1758 when the English optician John Dollond received a patent for a lens of crown glass cemented to one of flint glass. By 1900 Karl Zeiss (Germany) developed 80 glass types.

Additional reading: Derry. *Short history of technology;* p. 112–113, 592–593, 599.

Leukocyte
HISTOLOGY

While studying simple animal life under a microscope in 1883, Russian-born bacteriologist Ilya Ilich Mechnikov (France) observed that certain cells seemed to move to damaged areas where they would ingest small particles. Mechnikov demonstrated that these cells corresponded to white blood corpuscles in higher animals and were ingesting **bacteria.** Mechnikov shared the 1908 Nobel Prize in medicine and physiology with Paul Ehrlich (Germany) for his work on white blood corpuscles, now also known as leukocytes.

Additional reading: Magner. *A history of the life sciences;* p. 273–275.

Lever. *See* **Gear.**

Levodopa. *See* **L-Dopa.**

Leyden Jar. *See* **Electric Condenser.**

Librium
PHARMACOLOGY

In 1958 Leo Sternbach (U.S.) filed a patent on the first of the benzodiazepines, a new group of tranquilizers. Sternbach's drug was marketed as Librium when the FDA granted approval in 1960.

Additional reading: Sneader. *Drug discovery;* p. 184.

Lie Detector. *See* **Polygraph.**

Light
PHYSICS

For centuries scientists disagreed as to whether light consists of corpuscles (minute particles) or has the characteristics of a wave. The Dutch physicist and astronomer Christiaan Huygens presented a paper in 1678 in which he set forth his wave theory. This set him at odds with the corpuscular theory of the English scientist Sir Isaac Newton, stated around 1666. In that year Newton discovered that white light is made up of all colors. The wave theory received support in 1801 by the work of the English physicist Thomas Young. Measurement of length of light waves was made possible by the **interferometer.** The particle concept was furthered by the **quantum theory.** (See also Light Diffraction; Light Refraction; Light Velocity)

Additional reading: Asimov. *History of physics;* p. 300–305, 307.

Light Diffraction
PHYSICS

The ability of light to travel around small objects was discovered in 1818 by the French physicist Augustin Fresnel. This process, called diffraction, forms distinctive patterns, allowing wave lengths of light to be calculated. Diffraction gratings were made by the German physicist and optician Joseph von Fraunhofer around 1820. Much more precise gratings were made around 1880 by the American physicist Henry Rowland.

Additional reading: Asimov. *Asimov's biographical encyclopedia;* p. 303–307, 518–519.

Light Emitting Diode
SOLID-STATE PHYSICS

Light emitting diodes (LED) are **semiconductor** devices in which electrons subject to an applied voltage subsequently emit radiation as they fall to lower energy levels. The color of the radiation depends upon the energy loss. These devices are often used to light displays in computers and other electronic products. One team investigating this phenomenon was that of J. W. Allen and P. E. Gibbons (both English) in 1959. The first practical LED was invented in 1962 by Nick Holonyak, Jr. (U.S.). His device used gallium arsenide phosphide.

Additional reading: *Illustrated science and invention encyclopedia;* v. 12, p. 1628–1632.

Light Pen. *See* **Pattern Recognition.**

Light Refraction
PHYSICS

The change in the direction of travel of light as it passes from one medium to another is called refraction. The law governing this process was first developed in 1621 by the Dutch physicist Willebrord Snell. It was never published by him and was independently discovered by the French philosopher Rene Descartes in 1637. A related phenomenon was discovered by the Danish scientist Erasmus Bartholin in 1669 when he observed that objects viewed through transparent crystal (called Iceland spar) appeared double. He assumed the crystal caused light to be refracted in two different angles—double refraction. This later led the English physicist Thomas Young to theorize (1817) that light waves vibrated at right angles to the direction of the waves. (*See also* Polarized Light)

Additional reading: Asimov. *Asimov's biographical encyclopedia;* p. 113, 133–134, 269–271.

Light Velocity
PHYSICS

The speed of light has been the subject of studies for several centuries. The first fairly successful measurement was announced by the Danish astronomer Olaus Roemer in 1676, who calculated it at 132,000 miles per second. More accurate figures were obtained around 1849 by the French physicists Jean Foucault and Armand Fizeau. Calculations by the Scottish mathematician James C. Maxwell in 1864 were close to modern standards. One of the most accurate measurements (186,320 miles per second) was made in 1882 by the German-born physicist Albert Michelson (U.S.). He and the chemist Edward Morley (U.S.) collaborated in an experiment in 1887 proving the velocity of light in a vacuum to be a constant, the top speed at which any object can move (now accepted as 186,282 miles per second).

Additional reading: Asimov. *Asimov's biographical encyclopedia:* p. 155, 403, 456, 542.

Lighthouse
CIVIL ENGINEERING

Among early structures the lighthouse was an important example. One of the earliest was begun around 270 B.C. by the Greek architect Sostrastes of Cnidos. It was known as Pharos of Alexandria and was approximately 400 feet high. One of the seven wonders of the ancient world, this lighthouse lasted until the thirteenth century. Later examples were built by the Romans.

Additional reading: Clark. *Works of man;* p. 21–22, 1971.

Lightning Rod
ELECTRICITY

The phenomenon of lightning undoubtedly puzzled mankind for centuries. One of the first to study it was Benjamin Franklin (U.S.), who discovered around 1749 that lightning was a form of electricity. In 1753 he invented the lightning rod or lightning conductor, used for the purpose of conducting lightning harmlessly to the earth.

Additional reading: *Illustrated science and invention encyclopedia;* v. 10, p. 1370–1371.

Linotype Machine. *See* **Typesetting.**

Liquefaction of Gas — PHYSICS

Early studies of gas temperatures included those of the French physicist Jacques Charles, who discovered in 1787 that when a gas is cooled each degree of cooling causes its volume to contract about 1/273 of its volume at 0 degrees. In 1848 William Thomson, a Scottish physicist later to be known as Lord Kelvin, noted that at -273 degrees Centigrade the point of absolute zero would be reached, thus liquefying all gases at or near that point. Even before Kelvin's finding, the English physicist Michael Faraday in 1823 succeeded in liquefying gases such as carbon dioxide and chlorine.

Additional reading: Asimov. *Asimov's new guide to science;* p. 294–299.

Liquid Crystal. *See* **Crystallography.**

Lithium — CHEMISTRY

A soft, silver-white metal, lithium, atomic number 3, is the lightest in weight of all solid chemical elements. It was discovered in 1817 by the Swedish scientist John Arfwedson. The main sources are pegmatite ores and certain brines. It is used in various alloys, in storage batteries, and as a coolant for nuclear reactors. Chemical symbol is Li.

Additional reading: *Van Nostrand's scientific encyclopedia;* p. 1773–1774.

Lithium (Biological Aspects) — PHARMACOLOGY

An Australian psychiatrist named John Cade studied the affect of lithium on manic-depressive patients in the late 1940s. Feeling that manic-depressives might metabolize too much uric acid, he injected uric acid in the form of lithium salts and lithium carbonate into guinea pigs. In 1949 Cade published research results confirming his suspicions. Danish psychiatrist Mogens Schou continued Cade's line of research in the mid-1950s and was able to confirm the beneficial affects lithium had when treating symptoms of mania. Lithium was introduced to American psychiatric practice in the 1970s.

Additional reading: "20 discoveries that changed our lives"; p. 142.

Lithography — GRAPHIC ARTS

A widely used method of printing, lithography was discovered accidentally by the Czech writer Aloys Senefelder in 1796. While writing on a slab of limestone with a grease pencil he found that nonrelief printing would result. By wetting the surface of a stone after a greasy ink was applied, any design previously put on the stone would hold the ink and transfer it to a piece of paper. Among other improvements since then has been the process of electron beam lithography.

Additional reading: Feldman. *Scientists & inventors;* p. 102–103.

Liver Transplant — MEDICINE

The first liver transplant between humans was performed in 1963 by American physician Thomas E. Starzl. The recipient survived for 23 days. The problem of organ rejection remained the biggest obstacle to success as known immunosuppressant drugs were not effective. Liver transplants were still considered to be in the experimental stage well into the 1970's.

Additional reading: Robinson. *The miracle finders;* p. 81–84.

Lobotomy — MEDICINE

In 1891 Swiss psychiatrist Gottlieb Burckhardt performed a surgical technique in which he removed part of the brain surface in order to make patients become docile. Portuguese surgeon Antonio Egas Moniz took this further in 1935 when he bored two holes in a patient's forehead and destroyed the prefrontal lobes. The lobotomy, as it was called, was considered effective treatment for schizophrenics. Moniz was awarded a share of the 1949 Nobel Prize for medicine and physiology for developing this new area of medicine known as psychosurgery.

Additional reading: Valenstein. *Great and desperate cures;* p. 80–166.

Lock (Canal) — CIVIL ENGINEERING

A crude sort of lock was used on rivers in China around 70 A.D., consisting of a dam with a removable section through which a boat could be drawn. The removable part was called a flash lock. With the invention of the pound lock, in which operators could raise or lower the level of an enclosed section, the Chinese began use of this version around 1000 A.D. The first pound lock in Europe appeared in Holland in 1373. In 1495 the Italian inventor Leonardo da Vinci devised a form of pound lock having gates that formed a vee when shut. Most locks have rises less than 30 feet, but a dam in the U.S.S.R. on the Irtish River has a rise of 138 feet. Locks are important parts of the Panama Canal and the St. Lawrence Seaway.

Additional reading: *Illustrated science and invention encyclopedia;* v. 11, p. 13922-1394.

Lock (Security) — MECHANICAL ENGINEERING

One of the earliest forms of door lock was a wooden one made in Egypt around 2000 B.C. Specimens have been found in pyramids or pictured in sculptures. More complicated locks involving a key to be turned were made by the Romans around 50 B.C. In 1778 Robert Barron (England) invented the tumbler lock, and in 1784 Joseph Bramah (England) devised a cylindrical slotted key which depressed several springs, each a different distance. In 1865 Linus Yale, Jr. (U.S.), invented the pin-tumbler cylinder lock, a secure lock in which the serrations on the edge of the key pushed on spring-loaded pins. Since then the time lock and the combination lock were invented.

Additional reading: *Illustrated science and invention encyclopedia;* v. 11, p. 1389–1391.

Locomotive. *See* **Diesel Engine; Steam Locomotive**

Logarithm — MATHEMATICS

The processes of multiplication and division were greatly simplified by the discovery of logarithms, credited to the Scottish mathematician John Napier in 1614. Logarithms provide a way of expressing numbers by means of exponents used with a base number. Napier's system of "natural" logarithms used as a base a number approximately equal to 2.718. Soon thereafter the English mathematician Henry Briggs worked out a similar system but based instead on 10. It was simpler to use Briggs' "common" logarithms than Napier's method. In 1624 Briggs worked out logarithms for tens of thousands of numbers (to fourteen decimal places).

Additional reading: Asimov. *Asimov's biographical encyclopedia;* p. 96–97, 99–100.

Logic Circuit — COMPUTER SCIENCE

The advent of computers has brought great reliance on logic, particularly logic in algebraic form, which can be applied to electronic circuits. In 1847 the English logician George Boole proposed forms of algebra to express logic, now called **Boolean algebra**, summed up in a book he wrote in 1854. These ideas were used in designs for **calculating machines** and then computers as well as other digital circuits. Three logic devices are needed for a computer, known as the AND, OR, and NOT functions. One of the earliest logic circuits was created in 1919 by W. H. Eccles and F. W. Jordan (both English).

Additional reading: *Illustrated science and invention encyclopedia;* v. 11, p. 1403–1405.

Loom — TEXTILES

One of the earliest devices was the loom, which is known to have dated back to 4000 B.C., according to evidence found in Egyptian tombs. For centuries looms were built in about the same way, but in 1785 the English inventor Edmund Cartwright devised a steam-powered loom that was able to run mechanically, although in 1678 a French naval officer, DeGennes, had designed (but not carried out) plans for a mechanical loom. In 1822 the English inventor R. Roberts devised a loom that was entirely automatic; it was soon widely adopted. An improved automatic loom invented by J. H. Northrop (U.S.) is still the basis for improved versions in current use. Special looms were needed for textiles having patterns and designs. (*See also* Jacquard Loom)

Additional reading: *Illustrated science and invention encyclopedia;* v. 11, p. 1410–1413.

Loudspeaker — ACOUSTICS

Early phonographs used a mechanical vibration of the needle to actuate a flared horn speaker. With the development of transducers (which convert mechanical energy into electrical energy) came the dynamic loudspeaker, first patented in 1877 by the German inventor Ernst Wermer, then in 1898 by Sir Oliver Lodge (England). The first working model was invented in 1924 by C. W. Rice and E. W. Kelley (both U.S.). Movement of the needle sends electrical signals to the coils in the speaker, which move the speaker's diaphragm. The diaphragm movements create sound waves. Some newer versions are the ribbon speaker and the electrostatic speaker.

Additional reading: *Illustrated science and invention encyclopedia;* v. 11, p. 1414–1416.

LSD-25. *See* **Lysergic Acid Diethylamide.**

Lutetium — CHEMISTRY

Lutetium, atomic number 71, is a silver-gray metal which was discovered in 1907 by the French chemist Georges Urbain. Most lutetium is obtained by the processing of heavy rare earth metals. It is used in **semiconductors** and in **television** products. Chemical symbol is Lu.

Additional reading: *Van Nostrand's scientific encyclopedia;* p. 1789.

Lysergic Acid Diethylamide — ORGANIC CHEMISTRY

In 1938 Albert Hofmann (Switzerland) synthesized lysergic acid diethylamide, commonly known as LSD-25. By 1943 Hofmann realized that this compound had very strong psychotomimetic properties. It was manufactured by Sandoz in Switzerland.

Additional reading: Sneader. *Drug discovery;* p. 105–110.

Lysosome — CYTOLOGY

Belgian cytologist Christian Rene de Duve discovered and named lysosomes in 1955 while using an **electron microscope** to investigate cellular interiors. He found digestive organelles containing an **enzyme** that broke down ingested nutrients. Duve shared the 1974 Nobel Prize in medicine and physiology with Americans Albert Claude and George Emil Palade.

Additional reading: Dean. *Lysosomes;* p. 1–2.

Lysozyme — BIOCHEMISTRY

In 1922 Scottish bacteriologist Sir Alexander Fleming discovered a **protein** with bacteria-killing properties. Lysozyme, as the protein was called, existed in tears and mucus. Lysozyme was the first **enzyme** whose structure was completely analyzed in three dimensions; David Phillips (Wales) did so in 1965.

Additional reading: Judson. *The eighth day of creation;* p. 582–585.

M

Macadam Road — CIVIL ENGINEERING

The need for better roads in the late eighteenth century was met by the efforts of a Scottish surveyor, John L. McAdam. In 1823 his road-building methods, officially adopted for use in Great Britain, consisted of using two layers of crushed stone on a bed of compacted earth, raised in the center to facilitate drainage. The development of rubber automobile tires made necessary the use of a tar binder, resulting in the creation of **blacktop roads.**

Additional reading: Feldman. *Scientists & inventors;* p. 88–89.

Mach Number. *See* **Supersonic Flight.**

Machine Gun — ORDNANCE

One of the earliest forms of a machine gun was invented in the fourteenth century, consisting of a number of muskets mounted on a frame so that lighting black powder would set off all the barrels in turn. A gun devised by James Puckle (England) in 1718 fired 63 rounds in seven minutes. It was not until 1862 that the gun of Richard Gatling (U.S.) was successful in rapid firing, although it had to be operated manually by turning a crank. Then in 1884 the American inventor Hiram Maxim invented the first automatic machine gun. It used energy from the recoil of firing to produce gas that ejected the used cartridge, brought up the new cartridge, and fired it. Another model, devised in 1885 by John Browning (U.S.), was a gas-operated device.

Additional reading: *Illustrated science and invention encyclopedia;* v. 11, p. 1424–1426.

Machine Tool — MECHANICAL ENGINEERING

The invention of various machine tools (lathes, drilling machines, milling machines, etc.) all led to vastly improved ways of working metals. They brought greater speed and accuracy. As early as 1797 the English inventor Henry Maudslay built a slide rest for clamping the metal-cutting tool to a lathe. In 1810 he invented a lathe for cutting **screw** threads. He also designed planing, drilling, and milling machines, and trained many men who were to be engineers themselves. (*See also* Computer-Aided Processes)

Additional reading: Clark. *Works of man;* p. 81–83.

Magnesium — CHEMISTRY

Magnesium, a silver-white metal, atomic number 12, is malleable and ductile when heated. Discovered as an element in 1808 by the English chemist Sir Humphry Davy, it was not isolated in metallic form until 1828 by the French chemist Antoine Bussy. The most common method of commercial production is the electroyltic treatment of sea water, where it occurs plentifully. It is used extensively in alloys or as a structural metal where its light weight is important. Chemical symbol is Mg.

Additional reading: *Van Nostrand's scientific encyclopedia;* p. 1797–1799.

Magnetic Recording — ELECTRONICS

The concept of recording sound on magnetic tape originated in 1888 in an article by the English inventor Oberlin Smith, who proposed use of fabric strips containing iron filings. In 1898 the Danish inventor Valdemar Poulsen demonstrated magnetic recordings done on a steel wire. The first magnetic recording tape was invented in 1927 by J. A. O'Neill (U.S.), who used a paper tape, replaced in 1932 by plastic tape. The tape moves past an electromagnet which in response to electrical signals varies the frequency and strength of the magnetic field; this varies the magnetic pattern induced on the tape. Tapes are also used with **videocassette recorders.** The first audiocassette was demonstrated in 1963, produced by the Philips Company in Holland.

Additional reading: *Illustrated science and invention encyclopedia;* v. 11, p. 1431–1432.

Magnetic Reversal — GEOPHYSICS

As early as 1908 a French physicist, Bernard Brunhes, noted that some rocks had a magnetization in reverse of the earth's magnetic field. An English geologist, D. P. McKenzie, wrote a paper in 1967 with R. L. Parker (U.S.) in which he supported the concept of **plate tectonics** and also geomagnetic reversals. Studies in the 1960s had indicated that at intervals of 100,000 to 50 million years the whole magnetic field of the earth had repeatedly changed directions, taking place in a few thousand years, which is very quick in geological terms. The paper noted magnetic "stripes" in the sea floor in which different layers of rock had reversed magnetic properties. Many geologists seem to support this concept as part of the process of plate tectonics.

Additional reading: Gribbin. *Our changing planet;* p. 59–63.

Magnetism
PHYSICS

Probably the first one to make a scientific study of magnetism was Queen Elizabeth I's court physician, William Gilbert (England). Around 1600 he concentrated chiefly on **terrestrial magnetism.** In the 1820s the English scientist Michael Faraday discovered magnetic lines of force while working with iron filings near a magnet. Others studied **electromagnetism.** Many studies of magnetic properties of materials have been made. One involved the temperature at which materials lose their magnetism. (*See also* Curie Point)

Additional reading: Asimov. *Asimov's new guide to science;* p. 224–225.

Magneto. *See* **Automobile Ignition System.**

Magnetohydrodynamics. *See* **Plasma.**

Magnetosphere. *See* **Van Allen Belt.**

Magnetostrictive Effect
SOLID-STATE PHYSICS

Some materials when placed in a changing magnetic field will change their length at a rate twice that of the magnetic field since it is independent of the direction of the field. The reverse effect also occurs. Magnetostriction was discovered in 1847 by James Joule (England); other studies were made around 1887 by the Japanese physicist Hantaru Nagaoka. Around 1914 George Pierce (U.S.) studied both the **piezoelectric effect** and magnetostriction. This effect is often used to create sound waves in the range known as **ultrasonics.** The device used to create or detect such waves is called a **transducer.** This effect is closely related to electrostriction, in which changing electric fields affect the length of dielectric materials.

Additional reading: *Illustrated science and invention encyclopedia;* v. 19, p. 2516–2517.

Manganese
CHEMISTRY

Manganese, atomic number 25, is a brittle, silver-white metal which was discovered by the Swedish chemist Karl W. Scheele in 1774. Most commercial manganese in the United States is imported, with the Soviet Union and South Africa as leading sources. Its chief uses are in industrial chemicals, in certain alloys, and as a deoxidizer in molten steel, where much of the manganese is recovered. Chemical symbol is Mn.

Additional reading: *Van Nostrand's scientific encyclopedia;* p. 1816–1818.

Manned Spaceflight
AEROSPACE ENGINEERING

The space program of the United States to date has favored the use of manned space vehicles when possible rather than **space probes.** After trailing the Soviet programs with their early successes, such as putting the first man into space (Yuri Gagarin in April 1961), the Americans were finally first in one mission, landing men on the moon in 1969 (Colonel Edwin Aldrin and Neil Armstrong). The Apollo project was a great success, involving several trips of astronauts to the moon's surface. In recent years the space shuttle project has used a craft which flew into space and returned to earth for another trip. The tragic accident of the U.S. space shuttle in 1986 has raised questions about the extent of dependence upon manned flight versus space probes. (*See also* Space Station)

Additional reading: Baker. *Conquest;* p. 19–84, 142–169.

Manometer
PHYSICS

In its simplest form a manometer is a U-shaped glass tube filled with a liquid (usually mercury). It is used to measure the pressure of gases. The difference of the levels in the two limbs of the tube is proportional to the amount of pressure. The first version was invented in 1661 by the Dutch physicist Christiaan Huygens. To measure very low pressure a type of manometer known as a McLeod gauge may be used. It uses two thin capillary tubes in which actual measurements are made, since the low pressures require delicate measurement techniques.

Additional reading: *Illustrated science and invention encyclopedia;* v. 11, p. 1443–1444.

Map Making
MAPPING

Maps have been known since earliest times; the oldest world map surviving is Babylonian and dates from around 500 B.C. Early maps were drawn on stone, wood, bone, and leather, as well as on many types of paper. Ancient Greek and Roman maps are known to have existed, but few have survived. A few medieval maps made around 800 A.D. have been located. One of the most famous maps was the world map made by the Belgian cartographer Gerardus Mercator (Gerard Kremer) in 1569. It is still being used for sea charts, being a projection in which a compass course can be drawn as a straight line.

Additional reading: Bagrow. *History of cartography;* p. 26, 31, 45, 118–119, 132–133.

Mars
ASTRONOMY

The movements of the planet Mars were carefully observed before telescopes were invented. The Danish astronomer Tycho Brahe kept excellent, detailed records around 1580, continued after his death by his assistant, the German astronomer Johann Kepler. The discovery of the inclination of its axis in 1781 and the ice caps at its poles in 1784 were later made by the German-born Sir William Herschel (England). Two satellites of Mars were sighted in 1877 by the American astronomer Asaph Hill. In that same year the Italian astronomer Giovanni Schiaparelli noted dark lines or channels (which unfortunately got translated as "canals"). United States space probes in the period 1965-1975 took hundreds of photographs, and two soft landings of probes sent back much surface data.

Additional reading: Asimov. *Asimov's new guide to science;* p. 120–126.

Marshalling Yard. *See* **Railroad Switchyard.**

Maser
ELECTRONICS

A highly precise timing device, which depends on the constant rate at which molecules can be made to vibrate at microwave frequencies. Because of the constancy of these vibrations, physicist Charles H. Townes (U.S.) saw that the phenomenon could control timing devices. He invented the first maser in 1953, although two Russian physicists, Nicolai G. Basov and Aleksandr M. Prochorov, also developed maser theory at that time (1953). The first continuous maser was invented by Nicolaas Bloembergen (Holland) in 1956. Related to the **atomic clock** and the **laser**, "Maser" is an acronym for "microwave amplification by stimulated emission of radiation."

Additional reading: Asimov. *Asimov's new guide to science;* p. 451–453.

Mass Conservation. *See* **Conservation of Mass; Energy and Mass.**

Mass Production
INDUSTRIAL ENGINEERING

The use of precisely designed tools and jigs to produce interchangeable parts led to what is generally called mass production, replacing the costly method of products individually built and fitted by craftsmen. The main credit for this development goes to the American inventor, Eli Whitney, who originated the method in 1797, while producing muskets for the United States government. It was not until 1801 that Whitney was able to demonstrate his method on a working basis. Since then modern production methods have gradually become common. Henry Ford (U.S.) popularized mass production with the introduction in 1908 of assembly lines to make his Model T automobile.

Additional reading: *Engineers and inventors;* p. 65–66, 156.

Match
MATERIALS

A type of match was known in China in 970 A.D. Around 1270 Marco Polo mentioned seeing matches on his trips. In 1570 sulfur matches were being used in England. Then in 1681 the chemist Robert Boyle (Ireland) invented sulfur and phosphorus-tipped matches. In 1827 the English inventor John Walker developed matches using potassium and antimony compounds, but they were difficult to strike. The use of white phosphorus in the late nineteenth century caused illnesses among factory workers, so red phosphorus, discovered in 1844 by the Swedish inventor Carl Lundstrom, replaced the white form. The use of a phosphorus sulfide improved the ease of igniting matches by friction. Match making is now a highly automated process.

Additional reading: *Illustrated science and invention encyclopedia;* v. 11, p. 1466–1468.

Matrix Mechanics. *See* **Uncertainty Principle.**

Matter Waves. *See* **Wave Mechanics.**

McLeod Gauge. *See* **Manometer.**

Measles Vaccine
IMMUNOLOGY

The measles vaccine was developed in 1954 by John Franklin Enders, an American virologist, and Thomas Peebles, an American pediatrician, for use as an active immunization against measles.

Additional reading: Robinson. *The miracle finders;* p. 47–49.

Meiosis
CYTOLOGY

In 1883 E. van Beneden, a Belgian zoologist, discovered the process of meiosis while studying a horse parasite, Ascaris, in hopes of learning about **chromosome** behavior. The process involves the halving of the number of chromosomes to create haploid cells preceding the point at which the **sperm** and egg join in **fertilization.** The German biologist, August Friedrich Leopold, understood the theoretical underpinnings when he stated in 1892 that the process of fertilization restores the original quantity of chromosomes.

Additional reading: Gribbin. *In search of the double helix;* p. 43–44.

Mendelevium
CHEMISTRY

A radioactive metal of the actinide series, mendelevium, atomic number 101, was discovered in 1955 by five American chemists, A. Ghiorso, B. G. Harvey, G. R. Choppin, S. G. Thompson, and Glenn T. Seaborg. It is produced by bombardment of heavy elements with various substances. No known use exists at this time. Chemical symbol is Md.

Additional reading: *Van Nostrand's scientific encyclopedia;* p. 1853–1854.

Meningitis Vaccine
IMMUNOLOGY

Americans Malcolm S. Artenstein, Irving Goldschneider, and Emil C. Gotschlich set about in the 1960s to find an effective inoculation against meningitis, an inflammation of the membranes covering the brain and spinal cord caused by bacteria. This was necessary because sulfa drugs had lost their effectiveness against meningitis. Gotschlich was able to isolate the substance in meningococcus that stimulated production of antibodies in all nonhuman animals. The research team made vaccines with that substance, and in 1968 were able to run successful tests.

Additional reading: Robinson. *The miracle finders;* p. 59–63.

Mental Test
PSYCHOLOGY

American psychologist James McKeen Cattell coined the term "mental test" in 1890. A mental test was designed to measure a person's abilities, and evaluated such factors as strength of grip, speed of arm movement, and reaction time for sound. (*See also* Intelligence Test)

Additional reading: Murray. *A history of Western psychology;* p. 325–330.

Mercury
CHEMISTRY

Mercury, the only metal to exist in liquid state at temperatures below the freezing point of water, is atomic number 80. This silver-white liquid was used in Egypt as early as 1500 B.C., being one of the first metals known. It is highly poisonous in each of its physical states. The most common source is from the processing of cinnabar, found chiefly in Italy and Spain. Its most important uses are in electrical operations, in industrial chemicals, in pharmaceuticals, and in pesticides/herbicides. Chemical symbol is Hg.

Additional reading: *Van Nostrand's scientific encyclopedia;* p. 1856–1858.

Mercury (Planet)
ASTRONOMY

Mercury is hard to observe because it is the planet closest to the sun. Around 1610 it was discovered that Mercury had phases like our moon, proving that it revolved around the sun, as Copernicus declared. In 1831 the French philosopher Pierre Gassendi measured how long it took Mercury to cross the face of the sun. In 1889 the Italian astronomer Giovanni Schiaparelli studied the markings on Mercury and estimated it took 88 days for one revolution of Mercury. Microwave reflections from Mercury in 1962 showed the rotation to be about 59 days, as calculated by the two American electric engineers Rolf Dyce and Gordon Pettingill. Since 1973 space probes have provided a wealth of information about the planet.

Additional reading: Asimov. *Asimov's new guide to science;* p. 114–116, 119.

Mercury Vapor Lamp. *See* **Fluorescent Lighting.**

Meson
NUCLEAR PHYSICS

Studies of **nuclear interaction** convinced the Japanese physicist Hideki Yukawa in 1935 that there must be a short-range force in order to keep the nucleus intact. The force could be caused by an exchange particle with a mass around 300 times the mass of an electron. In 1936 the American physicist Carl Anderson found evidence of a particle having a mass in the right range. Anderson suggested the name mesotron, but it was shortened to meson. In 1960 it was found to be only a heavy electron, not the particle sought. In 1947 the English physicist Cecil Powell discovered a heavier muon, which did fit the Yukawa specifications. Powell's meson was called the pi-meson and Anderson's the mu-meson or muon.

Additional reading: Asimov. *Asimov's biographical encyclopedia;* p. 797–798, 805–806.

Messenger-RNA
BIOCHEMISTRY

In 1961 Francois Jacob (France) and Jacques Lucien Monod (France) hypothesized the existence of a "messenger-RNA" that carried the **DNA** blueprint from the nucleus to the **ribosome,** the cytoplasmic site of **protein** formation. Jacob shared the 1965 Nobel Prize for medicine and physiology with Monod and another Frenchman, Andre Michael Lwoff, for research in the regulation of gene action. (*See also* RNA; Transfer-RNA)

Additional reading: Gribbin. *In search of the double helix;* p. 258–294.

Metric System
SCIENCE AND TECHNOLOGY

For centuries scientists and engineers had felt the need for a uniform system of weights and measures, and in 1670 such a system was proposed in France. Despite the awkwardness of having a different system in most countries, as well as having three systems of weight in England alone, no action was taken until the late eighteenth century. In 1791 a committee of the French Academy of Science made a report, proposing what they called the metric system. Names were given for multiples and fractions of basic units (supplemented by some larger and smaller units, adopted in 1958 by the International Committee on Weights and Measures). Not every country has officially adopted this system for general use, although scientists have universally used it.

Additional reading: Garard. *Invitation to chemistry;* p. 122–129.

Michaelis-Menten Equation
BIOCHEMISTRY

In 1913 German chemist Leonor Michaelis and his assistant Maud Lenora Menten (Germany) developed the Michaelis-Menten equation, describing the way in which an **enzyme** carried out its specific functions.

Additional reading: Asimov. *Asimov's new guide to science;* p. 573–577.

Microbalance. *See* Balance (Chemical).

Microchip
SOLID-STATE PHYSICS

The trend toward miniaturization of electronic devices resulted in the development around 1960 of **integrated circuits;** these greatly reduced the size of individual circuit elements, produced on a small chip of **silicon.** By the 1970s further reductions in size made it reasonable that these tiny chips, the size of a fingernail but holding thousands of components, should be called microchips. They have brought great reductions of size to all types of electronic devices, particularly to computers, whose size, increased speed, and greater memory have improved tremendously. The United States has been the initiator of many of these developments.

Additional reading: Clark. *Works of man;* p. 287–291.

Microcomputer
COMPUTER SCIENCE

The development of computer components in microminiature size, so that a **microchip** the size of a thumbnail could contain tens of thousands of transistors, made smaller computers a very logical next step beyond minicomputers. Around 1971 the established computer manufacturers lacked interest in making microcomputers, so young computer buffs such as Edward Roberts began to make crude personal computers. By the end of 1976 several small computer manufacturers were creating a new industry—making microcomputers for an avid set of customers. The first operating system was devised in 1973 by Gary Kildall, a computer professor. Then in 1975 William Gates and Paul Allen wrote the first BASIC program and sold it to Roberts. Soon many more programs and makes of microcomputers were created.

Additional reading: Freiberger. *Fire in the valley;* p. 27–53, 136–143.

Microfilm
GRAPHIC ARTS

The photographic reduction of an image is called microfilm. The first commercial application of microfilm came in 1925 when a bank clerk, G. L. McCarthy (U.S.), obtained a patent for a machine to photograph checks. In 1928 the Eastman Kodak Company (U.S.) produced the first commercial 16mm microfilm cameras. Flow type machines use long reels of film and can copy up to 600 checks per minute, reducing the image at a ratio of 20:1. Greater reduction of up to 150:1 is sometimes used. Microfilm can be in reels, or 4" x 6" pieces called microfiche (introduced from France around 1960) or on punched cards (called aperture cards). Now little used, microcards use 3" x 5" opaque cards containing images protected by a glossy surface.

Additional reading: *Illustrated science and invention encyclopedia;* v. 11, p. 1504–1506.

Micrometer Gauge
ENGINEERING

A device for precise measuring of thicknesses or diameters (usually of metals) is called a micrometer gauge. It is based on the principle that turning a screw advances the gauge spindle a known precise amount. There is no clear record of when the first one was invented although a type of micrometer for a telescope eyepiece was devised around 1636 by the English astronomer William Gascoigne. A gauge for measuring metals was invented around 1800 by Henry Maudslay, an English engineer, who was able to measure one ten-thousandth of an inch. By 1856 Sir Joseph Whitworth, an English engineer, had devised a micrometer that could measure a millionth of an inch. Through modern technologies even more precise measurements can be made today.

Additional reading: Clark. *Works of man; p. 82–83.*

Microphone
ACOUSTICS

The conversion of sound waves into electrical signals can be done with microphones. Various studies of the process and creating of crude models took place in the middle of the nineteenth century, but not until 1877 did the American inventors Emile Berliner and Thomas Alva Edison devise a carbon microphone, consisting of a pack of carbon granules whose electrical resistance varied with pressure on a metal diaphragm. That same year a moving coil microphone was separately invented by Charles Cuttris (U.S.) and Ernst Werner Siemens (Germany). It relied on inducing a current in a coil attached to the diaphragm. In 1923 the ribbon microphone was created by the German inventors W. H. Schottky and Erwin Gerlach. That year an improved crystal microphone was made by A. G. Dolbear (U.S.).

Additional reading: *Illustrated science and invention encyclopedia;* v. 11, p. 1508–1511.

Microprocessor. *See* Microchip.

Microscope
OPTICS

Dutch spectacle maker Zacharias Janssen devised the first compound microscope in 1590 by combining two double convex lenses in a tube. In approximately 1610 Galileo Galilei (Italy) used such an instrument to study insect anatomy. In 1680 Anton van Leeuwenhoek, a Dutch biologist, built a simple microscope that magnified objects up to 200 times and used it to look at yeast cells. Joseph Jackson Lister, an English optician, devised an achromatic microscope that eliminated color distortion in 1830. (*See also* Electron Microscope)

Additional reading: Asimov. *Asimov's new guide to science;* p. 571, 651–652.

Microtome
ENGINEERING

In the 1830s a Czech physiologist, Johannes Evangelista Purkinje, invented a knife that was a precursor to the microtome. C. Chevalier (France) introduced the term "microtome" in 1839. In 1866 Wilhelm His, a Swiss anatomist interested in histology, invented the first useful microtome for preparing thinly sliced tissues suitable for microscopical study. He announced his invention in 1870.

Additional reading: Bracegirdle. *A history of microtechnique;* p. 117–134.

Microwaves. *See* Radio.

Middlings. *See* Grain Elevator.

Milk Condensation
AGRICULTURE

In 1856 Gail Borden (U.S.) developed a process for reducing the water content of milk and began canning condensed milk. The product became widely used when he was awarded a contract to supply the Union army with condensed milk during the Civil War. Borden built a milk condensation plant in Connecticut.

Additional reading: Schlebecker. *Whereby we thrive;* p. 128.

Milking Machine
AGRICULTURE

The first milking machine was introduced in the United States in 1905, followed by the first truly efficient machine in 1914. The milking machine saved between 16 and 58 man-hours a year per cow. Use of these machines spread during World War II and by 1945 approximately 365,000 were in use.

Additional reading: Schlebecker. *Whereby we thrive;* p. 254.

Miner's Lamp
MINING ENGINEERING

The deaths of miners due to poisonous gases brought the attention of the English chemist Sir Humphry Davy to the matter. Around 1815 he devised a lamp that would never rise in temperature above the ignition point of gases such as methane. Also the lamp's flame would change color when methane was present in any significant quantity.

Additional reading: Feldman. *Scientists & inventors;* p. 108–109.

Mineral Classification
MINERALOGY

The complex and abundant array of minerals caused confusion and inaccuracy among early mineralogists. Georgius Agricola (the Latin name taken by the German physician Georg Bauer) wrote a book in 1546 which served as a comprehensive system of mineral classification, the first such effort. In 1729 John Woodward (England) prepared a book constituting a massive catalog of English minerals and fossils. It became a standard reference work on their classification. A system of mineralogy based on crystal symmetry and chemical composition was developed in 1784 by Rene Just Hauy (France). In 1799 William Babington (England) created a classification scheme for minerals also based on their chemical composition. Other classification systems followed these pioneer efforts.

Additional reading: Faul. *It began with a stone;* p. 30, 56–57, 89, 117.

Mining
MINING ENGINEERING

Although informal mining must have taken place during the Stone Age, organized mining dates back to the Egyptians around 3000 B.C. Gold, silver, and copper were some of the most popular materials sought, usually by slave labor. This sort of manpower probably did the work in Roman mines, continuing past the collapse of the empire. The book *De Re Metallica* by Georgius Agricola (Germany) in 1556, was for centuries the best source of information. Modernization and mechanization began as early as 1698 with the steam-driven pump invented by Captain Thomas Savery, an English military engineer, for the purpose of pumping water from mines. Electric drills and mechanized hauling equipment followed, as did other improvements. (*See also* Coal Mining; Copper Mining; Gold Mining; Silver Mining)

Additional reading: *Illustrated science and invention encyclopedia;* v. 2, p. 176–177; v. 12, p. 1529–1534.

Missile
ORDNANCE

Missiles are now understood to be guided airborne vehicles, either remotely guided or self-directed to targets. The earliest ones were developed during World War II in 1944, particularly the German bombs powered by the V-2 rocket, designed by the German engineer Wernher von Braun (later U.S. citizen). By the 1950s he was designing **rockets** for American spacecraft. Some short-range missiles once used wire guidance, with electrical signals sent by the operator via the wires. More common is the use of radio or radar control. For the longest range, intercontinental ballistic missiles (ICBM) are used; they are more immune from **electronic countermeasures** because of using **inertial guidance** systems. (*See also* Laser; Nuclear Weapon)

Additional reading: Friedman. *Advanced technology warfare;* p. 45–47, 56–67, 84–89.

Mitochondria
CYTOLOGY

In 1890 Richard Altmann (Switzerland) reported that he had found bioblasts within cells and that they were elementary organisms living as intracellular symbionts. In 1898 German biologist C. Benda gave these organisms the name "mitochondria," meaning "threads of cartilage," which he mistakenly thought them to be.

Additional reading: Asimov. *Asimov's new guide to science;* p. 578–582.

Mitosis
CYTOLOGY

A. Schneider, a German cytologist, was the first to observe visible changes in the **cell nucleus** during cell division in 1873. By 1879 Walther Flemming (Germany) had studied and named all stages of this process wherein diploid cells are created. In 1882 he assigned the term "mitosis" to this process and published a landmark paper entitled "Cell Substance, Nucleus, and Cell Division." (*See also* Plant Cytology)

Additional reading: Portugal. *A century of DNA;* p. 37–40.

Modulus of Elasticity
PHYSICS

The measurement of a body's **elasticity** depends upon its shape, in which stress (the force per unit area) and strain (the ratio of the added dimensional increase to the original dimension) must be computed. For a given body several types of elasticity can be measured. A longitudinal force involves the stretch modulus, or Young's modulus, discovered by the English physicist Thomas Young around 1809. A bending force involves the shear modulus, or torsion modulus. When a change in volume is involved, the bulk modulus is used.

Additional reading: *Illustrated science and invention encyclopedia;* v. 7, p. 843–844.

Moho. *See* Mohorovicic Discontinuity.

Mohorovicic Discontinuity
GEOLOGY

There is a distinct boundary between the outer crust of the earth and the hotter plastic material making up the next layer (called the mantle). The boundary was discovered in 1909 by the Croatian geophysicist Andruya Mohorovicic, who made the discovery while studying the effects of a Balkan earthquake. It is often called Moho for short. The boundary varies in depth, being only eight to ten miles below the sea level in some places. An attempt to drill into the boundary was made in the 1960s but soon was abandoned due to a lack of funds.

Additional reading: Hurlbut. *The planet we live on;* p. 315.

Molecular Biology
BIOLOGY

William Thomas Astbury, an English physical biochemist, coined the term "molecular biology" in the late 1930s to describe the study of the physicochemical properties of molecules found in cells. Astbury was researching the structure of **protein** molecules in particular.

Additional reading: Medawar. *Aristotle to zoos;* p. 186–189.

Molecular Structure
CHEMISTRY

In 1939 American chemist Linus Carl Pauling introduced dramatic new information about the structure of molecules. Pauling applied quantum mechanics to develop a theory which stated that electrons interacted in pairs to form a stabler and less energetic system together than either had been separately. Pauling's theory was published in his book entitled *The Nature of the Chemical Bond.* He applied his ideas to living tissues in 1951 and suggested that **protein** molecules were arranged in helices. He was awarded the 1954 Nobel Prize in chemistry for his contributions to the understanding of molecular structure.

Additional reading: Asimov. *Asimov's new guide to science;* p. 507–550.

Molybdenum
CHEMISTRY

A tough, malleable silver-white metal, molybdenum, atomic number 42, was first identified by the Swedish chemist Karl W. Scheele in 1778. The main commercial source is the treatment of such minerals as molybdenite, powellite, and wulfenite. The largest producer is in Colorado. It is chiefly used as an additive to various types of steel and in other alloys where high temperature properties are important. Chemical symbol is Mo.

Additional reading: *Van Nostrand's scientific encyclopedia;* p. 1911–1913.

Monoclonal Antibody
IMMUNOLOGY

Cesar Milstein (England) and Georges J. F. Kohler (England) developed a method for creating monoclonal antibodies in 1975. Milstein and Kohler fused a myeloma cell from a mouse cancer with a white blood cell capable of making antibodies. This process resulted in a hybridoma able to produce an **antibody** that acted specifically against an invader cell. Milstein and Kohler received a share of the 1984 Nobel Prize in medicine and physiology for their research on monoclonal antibodies.

Additional reading: Sikora. *Monoclonal antibodies;* p. 1–12.

Monorail System MECHANICAL ENGINEERING

Railroads with one rail instead of two, with cars suspended beneath, are called monorails. The first one was patented in 1821 by the English engineer Henry Palmer. There was a delay of only three years before a line was built for it along the Thames. The early cars were horse-drawn but in 1869 the English engineer J. L. Hadden used steam power for a system in Syria. In 1894 an electric-powered line was built in France using the design of the framework created by the French inventor C. Lartique around 1850. Gyroscopes were used in a line invented in 1903 by the Irish engineer Louis Brennon; two gyroscopes kept the cars upright, holding 50 people. In 1930 the Scottish inventor George Bennie devised a system in which airplane propellers powered the units.

Additional reading: *Illustrated science and invention encyclopedia;* v. 12, p. 1551–1554.

Monotype Machine. *See* **Typesetting.**

Moon ASTRONOMY

The nature of the moon remained largely a mystery until the invention of the telescope and its improvement by the Italian astronomer Galileo Galilei in 1610; he discovered that the moon's surface was covered with irregularities or craters. This was the first study of the moon, although ancient philosophers had speculated about it. The Greek explorer Pytheas theorized around 300 B.C. that the moon was responsible for the tides, and about 130 B.C. the Greek astronomer Hipparchus correctly estimated the distance to the moon. In 1757 the French mathematician Alexis Clairaut estimated the mass of the moon. In modern times moon probes began in 1959, and in 1969 Neil Armstrong (U.S.) was the first man to land and walk on the moon.

Additional reading: Asimov. *Asimov's new guide to science;* p. 104–114.

Morphine PHARMACOLOGY

In 1805 Friedrich Wilhelm Adam Ferdinand Serturner (Germany) isolated morphine, the active ingredient in **opium**, and called it "principium somniferum." Serturner published a report in 1817 in which he referred to this alkaloid as "morphium," a name changed to "morphine" that year by French chemist Joseph Gay-Lussac so that it would conform to his standard nomenclature for naming plant bases. Serturner is credited with founding alkaloid chemistry for his work with morphine. The use of morphine for medical purposes was advanced during the 1800s by French physiologist Francois Magendie. In 1923 John Gulland (England) and Sir Robert Robinson (England) determined the chemical structure of morphine. (*See also* Heroin)

Additional reading: Sneader. *Drug discovery;* p. 6–7, 72–75.

Morse Code. *See* **Communications Code.**

Motion PHYSICS

The laws which govern the motion of bodies were discovered by the English scientist Sir Isaac Newton around 1665. His laws dealt with inertia, acceleration of bodies and the relation of actions to the resulting reactions. The laws made possible the calculation of the movement of the planets, originally set forth by Kepler. (*See also* Planetary Motion)

Additional reading: Asimov. *Asimov's biographical encyclopedia;* p. 148–154.

Motion Pictures GRAPHIC ARTS

Efforts to produce motion pictures date back to the early 1800s when mechanical machines flashed a series of photographs before a viewer, giving the appearance of motion. A more complicated device was invented by Englishman Eadweard Muybridge (originally Edward Muggeridge) around 1870. His zoopraxiscopes used a light source to illustrate in turn a series of drawings on a screen. In 1889 Thomas Alva Edison (U.S.) developed his "kinetograph," using a flexible celluloid film developed that year for him by George Eastman (U.S.). The work of Edison was soon outdistanced by Auguste and Louis Lumiere in Paris, who developed in 1895 their "cinematographe" The first sound picture was shown in 1927 and the first wide screen (Cinerama) came in 1952. Color film was invented around 1932.

Additional reading: Feldman. *Scientists & inventors;* p. 206–207, 222–223, 242–243.

Motor. *See* **Electric Motor.**

Motorcycle MECHANICAL ENGINEERING

While striving to invent a gasoline-driven automobile in 1885, the German engineers Gottlieb Daimler and Wilhelm Maybach mounted a gasoline engine on a wood bicyle and tested it successfully, thus creating the first motorcycle. In 1889 the French inventor Felix Millet attached a five-cylinder engine to a tricycle, but it was considered too complicated to manufacture. In 1897 two French inventors, Eugene and Michel Werner, fastened a motor over the front wheel of a bicycle and found great success in sales of their front-wheel drive vehicle. By 1900 they mounted the motor in the lower loop of the frame, a safer design feature.

Additional reading: Giscard d'Estaing. *World almanac book of inventions;* p. 9–13.

Mower AGRICULTURE

The first mower was patented by Cyrenus Wheeler (U.S.) in 1856 and was used to cut hay and corn close to the ground without clogging and tearing the sod. It had two wheels and a flexible cutter bar.

Additional reading: Schlebecker. *Whereby we thrive;* p. 122–123.

Mu-meson. *See* **Meson.**

Multimeter. *See* **Voltmeter.**

Muon. *See* **Meson.**

Muriatic Acid. *See* **Hydrochloric Acid.**

Mutation GENETICS

The term "mutation" was adopted in 1901 by Hugo de Vries (Holland) to describe sudden and spontaneous changes in the hereditary material of Oenothera. De Vries claimed that these mutations accounted for key evolutionary steps in the creation of new species, as opposed to Charles Darwin's (England) theory (1859) that involved slight variations. In 1907 Thomas Hunt Morgan (U.S.) discovered many cases of mutation among generations of Drosophila, and thus expanded de Vries' theory to include the animal world. Hermann Joseph Muller (U.S.) was able to artificially induce mutations in Drosophila through application of ionizing radiations in 1927, and was awarded the 1946 Nobel Prize in medicine and physiology for this work. (*See also* Evolution)

Additional reading: Gribbin. *In search of the double helix;* p. 45–82.

N

Nail BUILDING CONSTRUCTION

For centuries nails were made by hand, later, they were cut from sheets of rolled iron. In 1786 a nail-making machine was invented by Ezekiel Reed (U.S.). A major improvement came in 1851 with the creation by Adolph Browne (U.S.) of a machine for cutting nails from a coil of wire. Today a typical factory can make 1,400 nails per second or 300 million per week.

Additional reading: *Illustrated science and invention encyclopedia;* v. 12, p. 1572–1573.

Natural Selection. *See* **Evolution.**

Nature and Nurture
PSYCHOLOGY

The phrase "nature and nurture" was coined in 1874 by British anthropologist Sir Francis Galton. He used it to express two major influences, heredity and environment, on the development of an individual's character, personality, and abilities.

Additional reading: Murray. *A history of Western psychology;* p. 232.

Navigation
NAVIGATION

The problems of navigating ships were greatly eased by the publication of nautical almanacs and similar handbooks by the American mathematician and navigator Nathaniel Bowditch, beginning in 1772. His books became world-renowned, covering such topics as celestial navigation, winds, currents, tides, etc. In time **electronic navigation** superseded some of the need for works such as his. (*See also* Oceanography)

Additional reading: Downs. *Landmarks in science;* chap. 50.

Neodymium
CHEMISTRY

Neodymium, atomic number 60, is a soft, silver-gray metal which was first identified in 1885 by Baron von Welsbach, an Austrian chemist. It is quite plentiful, with commercial production centered in the treatment of bastnasite and monazite. It is used chiefly in **lasers**, capacitors, and as an additive in various alloys. Chemical symbol is Nd.

Additional reading: *Van Nostrand's scientific encyclopedia;* p. 1964.

Neon
CHEMISTRY

Neon, an odorless, colorless gas, atomic number 10, is a member of the group of noble gases. It was first discovered in 1898 by the Scottish chemist Sir William Ramsay and the English chemist Morris Travers. It is a constituent of the atmosphere to the extent that the best commercial sources for neon are plants engaged in the separation of oxygen and nitrogen from the air. Chief uses include advertising signs, street lamps, and various electrical devices. Chemical symbol is Ne.

Additional reading: *Van Nostrand's scientific encyclopedia;* p. 1965.

Neon Lighting. *See* **Fluorescent Lighting.**

Neosalvarsan. *See* **Arsphenamine.**

Neptune
ASTRONOMY

Before Neptune was discovered at least one astronomer observed it and recorded it as a star; namely, the French astronomer Joseph Lalande in 1801. In 1846 an English mathematician, John Couch Adams, calculated the irregularities observed in the orbit of Uranus and predicted where the unknown planet should be. He was unable to get an astronomer to investigate, unlike the French astronomer Urbain Leverrier, who made essentially the same calculations and was fortunate in getting the German astronomer Johann Galle to check the region. Within an hour the planet was located! In 1846 a satellite of Neptune was discovered, but not until 1939 did the Dutch-born astronomer Gerard Kuiper (U.S.) discover the second satellite. Space probes are scheduled to examine Neptune in 1989.

Additional reading: Asimov. *Asimov's new guide to science;* p. 142–144.

Neptunism
GEOLOGY

The theory that all rocks were formed in or by water was formally proposed in 1786 by the German mineralogist Abraham Gottlob Werner. He stated that a universal sea had once covered the entire earth. Others preceding him had also accepted this belief; namely, the French naturalist Jean Etienne Guettard (around 1751) and Nicholas Desmarest (a French geologist) in 1763. As late as 1850 the Scottish geologist Robert Jameson was still convinced that neptunism was a correct interpretation. Eventually it was abandoned by all geologists, one of the last being the American geologist Thomas Sterry Hunt, who lived until 1892.

Additional reading: Faul. *It began with a stone;* p. 85, 86, 88, 111–112.

Neptunium
CHEMISTRY

A radioactive metal of the actinide series, neptunium, atomic number 93, was first produced in 1940 by the U.S. physicists Edwin McMillan and Philip Abelson by bombarding uranium with neutrons. There is no known use for the element at this time, although alloys of such substances as iridium and platinum containing neptunium have been produced. Chemical symbol is Np.

Additional reading: *Van Nostrand's scientific encyclopedia;* p. 1966–1977.

Nerve Cell. *See* **Neuron.**

Nerve Growth Factor
BIOCHEMISTRY

During the 1950s Rita Levi-Montalcini (U.S.) and Stanley Cohen (U.S.) isolated and characterized nerve growth factor (NGF), adding to the understanding of factors that control cell growth. Levi-Montalcini and Cohen were awarded the 1986 Nobel Prize in medicine and physiology for their discoveries. In 1970 Ralph Bradshaw (U.S.), Ruth Angeletti (U.S.), and William Frazier (U.S.) determined the complete protein sequence of NGF.

Additional reading: Marx. "The 1986 Nobel Prize for physiology or medicine"; p. 543–544.

Nervous System. *See* **Neurology.**

Net
MATERIALS

Early civilizations must have made some sort of nets for fishing and catching game. Small fragments of nets believed to date from about 5000 B.C. have been recovered from European peat deposits. Nets of that approximate vintage have also been found in Egypt, where materials of that sort can be well preserved in the dry soil.

Additional reading: *Illustrated science and invention encyclopedia;* v. 20, Section 4.

Neurology
MEDICINE

Neurology began in the 1760s with the publication of Swiss physiologist Albrecht von Haller's eight-volume textbook on human physiology in which he rejected popular notions regarding nerves in favor of information he obtained experimentally. He showed, for example, that stimulus to a nerve produced a sharp muscle contraction and thereby established that nervous stimulation controlled muscle movement. Haller also demonstrated that nerves channel and carry impulses that produce sensation in tissues.

Additional reading: Asimov. *A short history of biology;* p. 116–119.

Neuron
CYTOLOGY

Rudolf Albert von Koelliker, a Swiss anatomist and physiologist, showed in 1849 that nerve fibers were elongated portions of cells. Koelliker's work proved to be a forerunner of Santiago Ramon y Cajal's (Spain) "neuron theory," which asserted that the nervous system consisted solely of nerve cells and their processes. Ramon y Cajal developed this theory in 1889 when he identified neurons as the functional cells of the brain. These cells were actually named "neurons" in 1891 by German anatomist Heinrich Wilhelm Gottfried von Waldeyer-Hartz. Ramon y Cajal shared the 1906 Nobel Prize in medicine and physiology with Camillo Golgi (Italy).

Additional reading: Asimov. *Asimov's new guide to science;* p. 816–833.

Neutrino
PARTICLE PHYSICS

The emission of **beta particles** from a radioactive nucleus puzzled many scientists because the particles did not carry as much energy as expected. In 1931 the Austrian-born Wolfgang Pauli (U.S.) theorized that another particle carrying the extra energy was emitted along with the beta particle. In 1932 the Italian-born physicist Enrico Fermi (U.S.) gave it the name of neutrino. In 1956 the American physicist Raymond R. Davis used a large tank of chlorine-containing compound to capture neutrinos emitted from the sun. He did so, although not in the quantity expected. (*See also* Antineutrino)

Additional reading: Asimov. *Asimov's new guide to science;* p. 348–352.

Neutron
NUCLEAR PHYSICS

The existence of an unknown radiation released from nuclei of certain atoms was reported in 1930 by the German physicists Walter Bothe and Herbert Becker. In 1932 two French physicists, Frederic Joliot and Irene Joliot-Curie, reported using the radiation to knock protons from a target. In that same year the English physicist Sir James Chadwick concluded that a new particle, rather than radiation, caused the effects noted. Because it had a neutral electrical charge, it was named the neutron. This new particle explained some of the mysteries of the structure of nuclei of the elements. Neutrons and **protons** are now sometimes called nucleons.

Additional reading: Asimov. *Asimov's new guide to science;* p. 328–330.

Neutron Star
ASTRONOMY

Neutron stars are 10 million to 1 billion times as dense as a **white dwarf star,** with a core temperature of about 1 billion degrees. Their existence was predicted in 1934 by the Swiss astronomer Fritz Zwicky, and in 1939 the American physicist Robert Oppenheimer proposed a theoretical model of a neutron star. One was discovered in 1967 by the English astronomer Anthony Hewish and a graduate student Jocelyn Bell while observing radio signals from the stars. Hewish called them pulsars and by the 1980s 400 pulsars were found. In 1968 the American astronomer Thomas Gold theorized that the pulsating sources were neutron stars, which are small and very dense. This seems to be generally accepted. Pulsars emit not only radio waves but also **X-rays** and ultraviolet and visible radiation.

Additional reading: Buedeler. *The fascinating universe;* p. 224–226.

Niacin
BIOCHEMISTRY

In 1928 Joseph Goldberger, the Austrian-born physicist (U.S.), and George Wheeler (U.S.) discovered that a heated yeast extract cured blacktongue, a canine disease similar to pellagra in humans. The extract turned out to be niacin.

Additional reading: Asimov. *Asimov's new guide to science;* p. 701–714.

Nickel
CHEMISTRY

Nickel, a malleable, hard, silver-white metal, atomic number 28, has been used for thousands of years as evidenced by ancient coins containing alloys which consisted of up to 25% nickel, later known as German silver. Nickel was first isolated in 1751 by the Swedish chemist Axel F. Cronstedt. It is found in several minerals, particularly pentlandite. Nickel is used in many alloys, with more than half the nickel produced used in stainless steel. Chemical symbol is Ni.

Additional reading: *Van Nostrand's scientific encyclopedia;* p. 1975–1978.

Nicotinic Acid. *See* **Niacin.**

Niobium
CHEMISTRY

Niobium, atomic number 41, is a ductile, bluish-tinged metal. It was first identified in 1801 by the English chemist Charles Hatchett. It is closely related to tantalum; both are found in such minerals as columbite and pyrochlore. Its chief use is in various alloys and in semiconductor research. Chemical symbol is Nb.

Additional reading: *Van Nostrand's scientific encyclopedia;* p. 1979–1980.

Nitric Acid
CHEMISTRY

The first known description of the preparation of nitric acid was a Latin treatise written around 800 A.D. by the Arabian alchemist Geber. After that, Oriental and European chemists described methods of producing it. In 1785 an improved method of producing **sulfuric acid** devised by Henry Cavendish (England) led to a cheaper method of producing nitric acid that used sodium nitrate and sulfuric acid. Most commercial production involves oxidizing ammonia using a platinum catalyst, a process devised by the Latvian-born Wilhelm Ostwald (Germany) around 1905.

Additional reading: *Van Nostrand's scientific encyclopedia;* p. 1981.

Nitrogen
CHEMISTRY

An important constituent of air, nitrogen, atomic number 7, is a colorless, odorless, and tasteless gas. It was discovered in 1772 by the Scottish chemist Daniel Rutherford. The main commercial source is the liquefaction of air while producing pure oxygen. It is used in industrial cooling, in fertilizers, in industrial compounds, and in industrial processes like welding. Chemical symbol is N.

Additional reading: *Van Nostrand's scientific encyclopedia;* p. 1983–1986.

Nitrogen Fixation. *See* **Fertilizer; Plant Nutrition.**

Nitroglycerine
MATERIALS

In 1846 the Italian chemist Ascanio Sobrero produced the dangerous and powerful explosive nitroglycerine. The Swedish inventors Emil and Alfred Nobel experimented with it, with Emil losing his life in an explosion in 1864. Alfred went on to invent **dynamite.**

Additional reading: Giscard d'Estaing. *World almanac book of inventions;* p. 45.

Nitrous Oxide
DENTISTRY

Nitrous oxide was discovered by British chemist and physicist Sir Humphry Davy in 1800. After experimentation with self and animals, he suggested it might be used during surgery to dull pain. American dentist Horace Wells first demonstrated its use for teeth extractions in 1844. When Wells' first public demonstration failed due to procedural error, its use as an anesthetic was delayed for several years. In 1862 Gardner Quincy Colton (U.S.) used it successfully, thereby gaining nitrous oxide a degree of popularity it had not previously obtained. Wells was proclaimed "discoverer of practical anesthesia" by the American Medical Association in 1870. (*See also* Anesthesia)

Additional reading: Ring. *Dentistry: an illustrated history;* p. 229–237.

Nobelium
CHEMISTRY

Nobelium, atomic number 102, is a radioactive metal of the actinide series. Its origin has been disputed, but strong evidence credits its identification in 1958 to American chemists A. Ghiorso, T. Sikkeland, J. R. Walton, and Glenn T. Seaborg. It is produced by bombardment of certain heavy elements. There are no known uses for it at present. Chemical symbol is No.

Additional reading: *Van Nostrand's scientific encyclopedia;* p. 1987.

Non-Euclidean Geometry
MATHEMATICS

During the nineteenth century a new viewpoint about the universe led to the development of a different version of the geometry made famous by Euclid. Nikolai Lobachevski (Russia) in 1829 and G. F. B. Riemann (Germany) in 1854 created what is known as non-Euclidean geometry. It proved to be more suitable for problems involving astronomy and relativity.

Additional reading: Asimov. *Asimov's biographical encyclopedia;* p. 325–326; 441–442.

Novocain
DENTISTRY

The development of Novocain, trade name for procaine hydrochloride, in 1904 as a local anesthetic is attributed to German chemist Alfred Einhorn. Found to be much safer than **cocaine**, Novocain revolutionized the practice of dentistry by eliminating pain from most treatments. (*See also* Anesthesia)

Additional reading: Ring. *Dentistry: an illustrated history;* p. 308.

Nuclear Fallout
NUCLEONICS

One serious effect of nuclear devices is the fallout that would accompany an accident at a **nuclear reactor** or the inevitable fallout resulting from the explosion of a **nuclear weapon**. Fallout has been studied since the beginning of the Nuclear Age and has been seen as a matter of great concern by people in all countries. The reactor accidents at Three Mile Island in the United States and at Chernobyl in the U.S.S.R. are obvious examples. Both long-range and short-range effects are known. Most scientists warn of the effects of fallout, one of the first being the American chemist Linus Pauling, who has done so since around 1950. Only a few have taken the position of the American physicist Edward Teller, who stated around 1954 that fallout was not a serious problem.

Additional reading: Asimov. *Asimov's new guide to science;* p. 484–488.

Nuclear Fission
NUCLEAR PHYSICS

The splitting of the nucleus of an atom occurred accidentally in an experiment done in 1934 by the physicist Enrico Fermi (U.S.) when he bombarded **uranium** with **neutrons**. He assumed he had created a transuranium element, but the results puzzled the German physical chemist Otto Hahn and the Austrian-born physicist Lise Meitner (Sweden). In 1939 they stated that nuclear fission had taken place. Rapid developments occurred after that, particularly the discovery by Fermi and associates in 1942 that a nuclear chain reaction could be self-sustained in what was then called an atomic pile. (*See also* Nuclear Reactor; Nuclear Weapon)

Additional reading: Asimov. *Asimov's new guide to science;* p. 464–472. .

Nuclear Fusion
NUCLEAR PHYSICS

The process of nuclear fusion, in which certain atoms are combined, was seen since the 1930s as a source of great energy. In 1938 fusion was proposed by the German physicist Hans Bethe as the process that fuels the stars, making it a source of unlimited energy. One pound of hydrogen when caused to fuse would produce 35 million kilowatt hours of energy. The problem is that controlled nuclear fusion has yet to occur in a laboratory. The main cause for this is the need to operate at temperatures in the millions of degrees, one reason the processes are referred to as thermonuclear reactions. In the late 1940s some physicists realized that fusion could be caused by a uranium-fission bomb, which was subsequently done. (*See also* Plasma; Thermonuclear Weapon)

Additional reading: Asimov. *Asimov's new guide to science;* p. 474–476, 488, 501–503.

Nuclear Interaction
NUCLEAR PHYSICS

The attempts by physicists to explain the forces that affect atomic nuclei have brought forth many theories. Some sort of nuclear interaction between the various particles in the nucleus was felt to be necessary. In 1932 the German physicist Werner Heisenberg proposed that **protons** were held together by what he called exchange forces. An explanation of the nature of these forces came from the Japanese scientist Hideki Yukawa in 1935. He theorized that there must be an exchange particle in the nucleus that would have a mass intermediate between that of an **electron** and a proton. The subsequent discovery of such a particle, the **meson**, strengthened his case.

Additional reading: Asimov. *Asimov's new guide to science;* p. 352–354.

Nuclear Power Station. *See* Nuclear Reactor.

Nuclear Reactor
NUCLEONICS

Early in the preparation of the atomic bomb in the 1940s it was necessary to determine the conditions under which nuclear fission would be self-sustaining. The reactor was about 30 feet by 20 feet in size and weighed about 1,400 tons. Under the direction of the Italian-born Enrico Fermi (U.S.) a chain reaction was reached in 1942. After the war many uses were found for nuclear reactors, from nuclear-powered ships and submarines to nuclear power plants for generating electricity for public use. The first such plant in the United States began operating in 1958. Critics of such plants cite the accidents in Pennsylvania at Three Mile Island in 1979 and in Russia at Chernobyl in 1986 as reasons for closing such plants. Fears of **nuclear fallout** are also expressed.

Additional reading: Clark. *Works of man;* p. 298–308.

Nuclear Structure. *See* Atomic Theory; Names of specific subatomic particles and radiation.

Nuclear Transplantation. *See* Clone.

Nuclear Weapon
NUCLEONICS

Work on the atomic bomb in the United States began in 1941, authorized by President Franklin Roosevelt, and known as the Manhattan Project. Roosevelt is said to have been influenced by a letter written in 1939 by Albert Einstein (U.S.), who did so at the urging of colleagues concerned that Germany might develop such a weapon. One of the key scientists among the hundreds involved was J. Robert Oppenheimer (U.S.) under whose direction the actual bomb was constructed and completed in 1945. Other countries developed their own in a few years. Even more powerful than this type of bomb, a fission bomb, was the **thermonuclear weapon.** Much study has gone into the problems of **nuclear fallout** from these weapons. (*See also* Nuclear Fission; Nuclear Fusion)

Additional reading: *Asimov's new guide to science;* p. 468–474, 484–488.

Nucleic Acid
BIOCHEMISTRY

Nuclein, as it was first known, was discovered in 1869 by Swiss physician Johann Friedrich Miescher. In the 1880s, the German biochemist Albrecht Kossel was the first to scientifically study the structure of the nucleic acid molecule, and isolated a series of nitrogen-containing compounds which he named adenine, guanine, cytosine, and thymine, otherwise known as the purines and the pyrimidines. Kossel received the 1910 Nobel Prize for medicine and physiology for his work on **proteins** and nucleic acids. During the 1900s, Russian-born Phoebus Aaron Theodore Levene (U.S.) showed nucleic acid to be of two varieties, **RNA** and **DNA.**

Additional reading: Portugal. *A century of DNA;* p. 51–70.

Nucleon. *See* Neutron; Proton.

Nucleoprotein
BIOCHEMISTRY

Nucleoprotein was isolated in 1866 by Swiss physician Johann Friedrich Miescher. During the 1880s, German biochemist Albrecht Kossel showed that Miescher's substance had a **protein** portion and a nonprotein portion—**nucleic acid.** Based on its characteristics, Kossel named the substance "nucleoprotein."

Additional reading: Olby. *The path to the double helix;* p. 97–121.

Numeration System MATHEMATICS
The notation system used today is only one of several systems used in different eras and countries. Our present system, called Arabic numerals, actually began in India among the Hindus around 800 A.D. Arab mathematicians learned the system, and a Persian book written about 800 reached Europe (after being translated into Latin) around 1200, possibly helped by the Italian mathematician Leonardo of Pisa. The printing of books helped popularize the Hindu-Arabic system in Europe. This system is generally recognized as being superior to the others because of the use of position to indicate values and also the use of zero.
Additional reading: Asimov. *Asimov's new guide to science;* p. 854–857.

Numerical Control. *See* **Computer-Aided Processes.**

Nylon TEXTILES
Man-made materials in the textile world prior to 1935 were really chemically treated natural products, like **rayon.** In that year the American chemist Wallace Carothers invented nylon, a fiber entirely made from chemicals. He had long been involved in research on **polymers,** and he located one material that became, after stretching, stronger than silk. Nylon, which was resistant to grease and dirt, soon was used in a multitude of applications, ranging from parachutes to stockings. A variety of similar synthetic materials has subsequently been created.
Additional reading: Feldman. *Scientists & inventors;* p. 286–288.

O

Oar NAVAL ARCHITECTURE
The use of oars to power ships dates back to around 3000 B.C. in Egypt. Greek ships began to use more than one bank of oars; triremes had three rows of around 170 rowers. Invented around 700 B.C. by a Greek shipbuilder, Ameinokles of Corinth, the trireme as a fighting ship was used until around 1200 A.D. Around 500 B.C. some Greek ships had four and even five banks of oars, but they were more cumbersome to operate than the triremes. The length of the oars was around 14 feet.
Additional reading: Landels. *Engineering in the ancient world;* p. 139–146, 151–153.

Obstetric Forceps MEDICINE
Peter Chamberlen (England) invented obstetric forceps between 1572 and 1626. Since sensibilities were offended by the use of iron instruments during the birth process, Chamberlen was forced to keep his invention hidden. His secret was eventually uncovered, however, and in 1747 a publication describing his forceps was issued in Amsterdam.
Additional reading: Bennion. *Antique medical instruments;* p. 113–123.

Ocean Liner NAVAL ARCHITECTURE
Large passenger ships capable of crossing the Atlantic Ocean by steam power did not appear until 1838, when the *Sirius* made the crossing, relying mostly on its paddle wheels. The first propeller-driven ship to cross the Atlantic was the *Great Britain.* It had been designed by the British engineer Isambard Kingdom Brunel in 1843. It was also the largest ship to be built of iron. In 1887 the English inventor Sir Charles Parsons built the fastest ship of its time, powered by a steam-turbine engine. By 1905 this type of power was the most common for liners.
Additional reading: Feldman. *Scientists & inventors;* p. 136–137; 220–221.

Oceanography OCEANOGRAPHY
The need of those operating sailing vessels for accurate oceanographic data such as currents, winds, ocean depths, and meteorological phenomena has existed since earliest times. The first person to provide much of this information was the American oceanographer Matthew Maury, whose first book was published in 1836. By 1850 he had collected logs from a thousand captains on their observations, allowing Maury to issue updated charts. He did extensive work on ocean floor mapping and on the Gulf Stream. (*See also* Navigation)
Additional reading: Downs. *Landmarks in science;* chap. 56.

Odometer. *See* **Automobile Speedometer.**

Office Automation. *See* **Digital Computer; Word Processing.**

Ohm's Law ELECTRICITY
A formula was devised in 1827 by the German physicist Georg Ohm regarding electric circuits. The law states that the amount of electric current flowing is directly proportional to the voltage and inversely proportional to the resistance.
Additional reading: Feldman. *Scientists & inventors;* p. 114–115.

Oil Lamp ILLUMINATING ENGINEERING
Archaeologists have discovered artifacts indicating that oil lamps were used as far back as 20,000 B.C. They were simple to make, consisting of a container for the oil (animal fat or oil) plus a wick (perhaps made of twisted plant fiber). By around 400 B.C. the lamp in Greece had the familiar "Aladdin lamp" shape. To get more light, some lamps had up to 20 wicks. During the Middle Ages candles were more commonly used because of widespread poverty, but by around 1800 lamps were again in general use. During the eighteenth century glass lamps began to appear, the first devised by the Swiss inventor Aime Argand in 1804. It had a glass chimney and a braided wick, which decreased smoke and produced a brighter light; in 1868 the Wells light was first to use oil under pressure.
Additional reading: *Illustrated science and invention encyclopedia;* v. 10, p. 1324–1325.

Oil Refining PETROLEUM ENGINEERING
Oil as it comes from oil wells needs to be refined, one reason being to separate certain constituents too valuable to burn, such as lubricating oil and bitumen. Around 1700 the processing of crude oil to produce what was then called lamp oil took place in several European countries. In 1857 James Miller Williams established a refinery in Canada. Over the years refining of oil became more and more commercialized, the most common process using **distillation.** In 1877 Samuel Van Syckel (U.S.) developed a multistep refining process. The process of reforming large crude oil molecules into smaller ones is known as **cracking.** Refineries can produce hundreds of different petrochemicals (derivatives of petroleum).
Additional reading: *Illustrated science and invention encyclopedia;* v. 12, p. 1615–1619.

Oil Well Drilling. *See* **Drilling Rig.**

Open-Hearth Process METALLURGY
For over a century the main method for working steel was the open-hearth process. It was invented in 1858 by the German-born engineer Sir William Siemens (England). As improved in 1884 by the French engineer Pierre-Emile Martin, the Siemens-Martin equipment consisted of a shallow hearth, open to the flames above it, with the molten metal on the bottom; addition of a pair of Siemens chambers at either end of the furnace improved combustion and increased temperature. The main advantage of the process was that any proportion of scrap metal or other raw materials could be used.
Additional reading: Feldman. *Scientists & inventors;* p. 166–167.

Ophthalmology
OPHTHALMOLOGY

Albrecht von Graefe (Germany) is known as the founder of modern ophthalmology because of his contributions during the mid-1800s. In particular, he gained international renown when he introduced the technique of iridectomy, wherein a portion of the iris is removed to treat glaucoma.

Additional reading: Gorin. *History of ophthalmology;* p. 132–139.

Ophthalmometer
MEDICINE

German physiologist and physicist Hermann Ludwig Ferdinand von Helmholtz invented the ophthalmometer in the 1850s to measure the eye's curvature. The ophthalmometer made it possible to quantify astigmatisms, eye chamber capacity, and the whole eye itself.

Additional reading: Bettmann. *A pictorial history of medicine;* p. 266–267.

Ophthalmoscope
MEDICINE

The ophthalmoscope was invented in 1851 by German physician and physicist Hermann Ludwig Ferdinand von Helmholtz for the purpose of examining the interior of an eye through its pupil.

Additional reading: Borin. *History of ophthalmology;* p. 127–131.

Opioid Receptor
BIOCHEMISTRY

In 1973 Candace Pert (U.S.) and Solomon Snyder (U.S.) demonstrated the existence of opioid receptors in rat brains and in nerve cells taken from guinea pig intestines. This discovery made it possible to study the action of morphine on the brain. (*See also* Endorphins; Enkephlins)

Additional reading: Crapo. *Hormones: the messengers of life;* p. 163–165.

Opium
PHARMACOLOGY

Swiss physician and alchemist Philippus Aureolus Paracelsus was the first to use tincture of opium for medical purposes, and named it "laudanum" in the 1500s. Opium, a product of the poppy, Papaver somniferum, was used in the 1600s by Thomas Sydenham, an English physician, as a pain reliever and a sleep inducer. (*See also* Heroin; Morphine)

Additional reading: Sneader. *Drug discovery;* p. 2–6.

Optical Glass. *See* Lens.

Organic Compound Synthesis
ORGANIC CHEMISTRY

Prior to the nineteenth century, chemists believed organic compounds ("the chemistry of life" in their terms) were quite distinct from inorganic compounds. This was disproved in 1828 when the German chemist Friedrich Wohler accidentally succeeded in producing urea, an organic compound, from inorganic substances. However, it was to be a century before this compound was manufactured commercially, but the process of organic synthesis had begun.

Additional reading: Feldman. *Scientists & inventors;* p. 130–131.

Organizer Effect (Biology)
EMBRYOLOGY

In 1918 a German zoologist, Hans Spemann, discovered that chemicals within an embryo acted as morphogenic stimuli and caused morphological differentiation to occur. Spemann named these chemicals "organizers." He received the 1935 Nobel Prize in medicine and physiology for this discovery.

Additional reading: Magner. *A history of the life sciences;* p. 208–210.

Oscilloscope
ELECTRONICS

The oscilloscope consists of a stream of electrons, shot from an electric gun against a fluorescent screen, used to display characteristics of electronic signals being studied. It was invented in 1897 by Karl Ferdinand Braun (Germany) and was based on the studies of **cathode rays** made by William Crookes. This device became the basis for the iconoscope, a tube created for use in **television.**

Additional reading: *Illustrated science and invention encyclopedia;* v. 4, p. 528–530; v. 12, p. 1646–1648.

Osmium
CHEMISTRY

Osmium, a hard bluish-white metal, atomic number 76, was discovered in 1804 by the English amateur scientist Smithson Tennant. Osmium, a member of the **platinum** group of metals, is usually obtained as part of the process of treating platinum. Its main use is in products requiring alloys having good resistance to wear, such as instrument pivots. Another use is in laboratory compounds, such as oxidizing agents. Chemical symbol is Os.

Additional reading: *Van Nostrand's scientific encyclopedia;* p. 2118–2119.

Osmosis
CHEMISTRY

The passage of a solution through a permeable membrane containing a more concentrated solution is called osmosis. It was discovered in 1748 by the French physicist Jean Antoine Nollet. He found that water passed through an animal membrane into a sugar solution. In 1826 the French chemist Henri Dutrochet found that the osmotic pressure was proportional to the degree of concentration. By applying pressure, the process is reversed, resulting in pure water on one side of the membrane. This is one way to desalinate brackish water.

Additional reading: *Illustrated science and invention encyclopedia;* v. 12, p. 1649–1650.

Osteopathy
MEDICINE

Andrew Taylor Still (U.S.) established osteopathy as a field in 1874. Still asserted that all diseases stem from vertebrate dislocations. He advocated treatments consisting of massage and physical manipulation and discouraged the use of drugs.

Additional reading: McGrew. *Encyclopedia of medical history;* p. 158–159.

Ovum
CYTOLOGY

In 1657 Dutch anatomist Regnier de Graaf described mammalian ovary structures now known as Graafian follicles. Graaf mistakenly believed he had identified the mammalian egg, and it was not until Russian-born Karl Ernst von Baer (Germany) opened such a follicle that he was corrected. Baer found a small yellow point that, upon examination under a microscope, turned out to be the actual egg. Baer's finding, which he published in 1827, clarified the nature of human development.

Additional reading: Magner. *A history of the life sciences;* p. 201–203.

Oxygen
CHEMISTRY

This important gas, atomic number 8, was discovered by the English chemist Joseph Priestly in 1774, although it was named by Antoine Lavoisier (France) in 1779. It had been previously discovered by 1772 by the Swedish chemist Karl W. Scheele, but his findings were not published until 1777, by which time Priestly's account had already appeared. The most economical method of production is to make liquid oxygen from air. Chief uses include industrial compounds, welding, and steel making.

Additional reading: Asimov. *Asimov's biographical encyclopedia;* p. 204–206; 222–226.

Ozone
METEOROLOGY

Ozone, which is found in greatest concentrations in the lower stratosphere (10 to 20 miles high), serves to absorb harmful **ultraviolet radiation** emitted by the sun. Ozone also heats the upper atmosphere by absorbing other components of solar radiation. Ozone, which is made up of three atoms of oxygen, was discovered in the atmosphere in 1913 by the French physicist Charles Fabry. It had been identified as a form of oxygen around 1840 by the Irish chemist Thomas Andrews.

Additional reading: Asimov. *Asimov's new guide to science;* p. 219–220.

P

Pacemaker
MEDICINE

Clarence Walton Lillehei, an American physician, built the first pacemaker in 1957. Lillehei's pacemaker was an electric unit that could be inserted in the patient's chest where it would give off an electrical jolt in order to regulate the pace of the heart beat.

Additional reading: Robinson. *The miracle finders;* p. 175–176.

Paint
MATERIALS

Probably the earliest use of paint for decoration was for the paintings in the caves of France and Spain, dating back to 15,000 B.C. Around 1500 B.C. the Egyptians used dyes to make blue and red pigments. In the Orient organic pigments were used as early as 6000 B.C., involving such materials as eggwhite and beeswax. The use of paint to protect surfaces did not occur much until the Middle Ages, and painters kept their formulas secret. Painter Jan Van Eyck helped bring oil paint into prominence around 1420 in Holland. White lead paint was in use by the seventh century. By the nineteenth century chemists and engineers were being employed in the manufacture of paint. In recent decades the use of solvents, plasticizers, adhesion promoters, and similar products have generally improved paint.

Additional reading: *Illustrated science and invention encyclopedia;* v. 12, p. 1664–1668.

Paleoecology
ECOLOGY

The science of paleoecology was founded in the 1840s by British naturalist Edward Forbes when he linked biological observations of sea bottom invertebrates to the prevalent geological ideas regarding the fossil record of such organisms and the environmental changes associated with them. The term "paleoecology" was suggested in 1916 by F.E. Clements (U.S.) to mean the study of past vegetation. (*See also* Fossil)

Additional reading: McIntosh. *Background of ecology;* p. 98–104.

Paleontology. *See* **Fossil; Paleoecology.**

Palladium
CHEMISTRY

A silver-white metal, palladium, atomic number 46, was discovered in 1803 by the English chemist William Wollaston. It is usually found in ores containing **platinum**, the extraction of which leads eventually to palladium. One of its chief uses is in catalytic converters for automobile exhaust systems, in laboratory equipment, and in industrial alloys. Chemical symbol is Pd.

Additional reading: *Van Nostrand's scientific encyclopedia;* p. 2134–2135.

Paper. *See* **Writing Material.**

Papyrus. *See* **Writing Material.**

Parachute
AEROSPACE ENGINEERING

The French inventor Jacques Garnerin made his first descent from a balloon in a parachute in 1797; he obtained a patent for his parachute in 1802. The first parachute jump from an airplane was made in 1912 by an American officer, a Captain Berry. During World War I balloonists and airmen used parachutes to avoid crashes. Recent years have seen new developments, such as a helmet which contains a barometric switch to trigger the parachute at a fixed altitude. Since 1939 nylon and other synthetic textiles have been used instead of cotton and silk.

Additional reading: *Illustrated science and invention encyclopedia;* v. 13, p. 1675–1677.

Parallel Processor. *See* **Supercomputer.**

Parchment. *See* **Writing Material.**

Parthenogenesis
ZOOLOGY

Anton van Leeuwenhoek, a Dutch biologist, was the first to observe parthenogenesis when, in the 1690s, he saw that female aphids were reproducing in the absence of males. Following this, Swiss naturalist Charles Bonnet discovered in the 1700s that aphids could reproduce parthenogenetically. Female eggs could develop without being fertilized by a male sperm. In the 1950s, Gregory Goodwin Pincus (U.S.) engineered the first fatherless mammalian birth when he took an ovum from a female rabbit, fertilized it with hormones and a salt solution, and implanted it in the rabbit's uterus. The egg developed into an embryo and fetus.

Additional reading: Magner. *A history of the life sciences;* p. 189–196.

Particle Accelerator
NUCLEONICS

The need to bombard elements in order to create subatomic particles led physicists to create particle accelerators. One of the earliest, developed in 1928 by the English physicists John Cockcroft and Ernest Walton, was named a voltage generator. A related device was the **Van de Graaf generator.** The linear accelerator was proposed in 1931 but never caught on. A successful device in the next few years was the **cyclotron.** Other successful types were the **betatron**, the **synchrotron**, the bevatron, and the **tevatron.** These devices brought important discoveries, including creating the **antiproton.**

Additional reading: Asimov. *Asimov's new guide to science;* p. 336–341.

Particle Spin
NUCLEAR PHYSICS

Almost every subatomic particle has a spin that creates a tiny magnetic field. The concept of spin was developed in 1925 by the Dutch physicists George Eugene Uhlenbeck and Samuel Abraham Goudsmit as a way to describe **electrons** in atoms. The units used for measuring spin are arranged so that spin is expressed either in whole numbers or half numbers. Those with whole-number spins are called **bosons** while those with half-number spins are called **fermions.** The early work on measurement of spin was carried out by the American physicists Otto Stern in 1920 and Isidor Isaac Rabi in 1933. The law of conservation of spin says that the total spin of particles involved in **nuclear reactions** remains constant.

Additional reading: Asimov. *Asimov's new guide to science;* p. 327–328, 342–343, 347, 349–350.

Pasteurization
MICROBIOLOGY

As a consequence of his work with **fermentation,** French chemist Louis Pasteur found that the process of heating a liquid killed organisms such as yeast cells, that could sour the substance. In 1856 he used this process of pasteurization to prevent excess fermentation in wine. In 1885 Pasteur first pasteurized milk.

Additional reading: Magner. *A history of the life sciences;* p. 237–257.

Pattern Recognition
COMPUTER SCIENCE

Pattern recognition, sometimes called character recognition, is the process of identifying a given pattern of data. The data might be alphanumeric (letters or numbers) or graphic symbols (such as pictures). Around 1956 in the United States, an early application was the use of magnetic ink on bank checks so they could be sorted quickly by computers. Beginning in the 1960s the U.S. postal service used optical scanners to sort addresses on pieces of mail. By the 1980s 97% accuracy was being achieved. Supermarkets and libraries, among others, commonly use light pens, developed in 1963 by I. E. Sutherland (U.S.) to read numbers on objects. Military applications abound, many of great complexity.

Additional reading: *Illustrated science and invention encyclopedia;* v. 5, p. 549–551; v. 14, p. 1822–1826.

Peltier Effect
PHYSICS

One of the phenomena known as **thermoelectricity,** the Peltier effect refers to the process in which an electric current flowing through a junction made up of two dissimilar metals will cause one junction to get hotter and the other colder. It was discovered in 1834 by the French physicist Jean Peltier. It is closely related to the **Seebeck effect** and has led to considerable use of **thermoelectric coolers.**

Additional reading: *Illustrated science and invention encyclopedia;* v. 18, p. 2404–2406.

Pen
GRAPHIC ARTS

Quills for writing were not in general use until the seventh century. The steel-nibbed pen was invented in 1748 by the French inventor Johann Janssen. In 1663 Samuel Pepys used a "reservoir pen." Two British patents involving pens were granted in 1809, and in 1833 the English inventor William Baddelley invented the plunger-filled fountain pen. The first free-flowing fountain pen was invented in 1884 by the American inventor L. E. Waterman. The first ball-point pen was invented in 1888 by the American inventor John Loud, but it was not popular. In 1938 two Hungarians, Ladislao Biro and his brother Georg, patented a ballpoint pen, which they developed in Argentina, where they had moved to escape the Nazis. Patented in Argentina in 1943, it soon became popular.

Additional reading: *Illustrated science and invention encyclopedia;* v. 8, p. 1059–1060.

Pendulum Clock. *See* Clock.

Penicillin
MICROBIOLOGY

An **antibiotic** produced by the mold Penicillium notatum that dissolves staphylococci microbes, penicillin was discovered by Alexander Fleming (Scotland) in 1928, ushering in the modern age of antibiotics. In 1940 the Australian-born pathologist Howard Walter Florey (England) and German-born biochemist Ernst Boris Chain (England) isolated penicillin, and in 1941 it was successfully used to treat human bacterial infection. Large scale production began in 1943. Fleming, Chain, and Florey shared the 1945 Nobel Prize in medicine and physiology for their work with penicillin.

Additional reading: MacFarlane. *Alexander Fleming: the man and the myth.*

Pepsin
BIOCHEMISTRY

German physiologist Theodor Schwann discovered and isolated pepsin in the human stomach in 1836. He demonstrated that this **enzyme** increased the stomach's ability to digest meat and named it "pepsin" from the Greek word meaning "to digest." In 1930 American biochemist John Howard Northrop crystallized pepsin. He shared the 1946 Nobel Prize in chemistry with James Batcheller Sumner (U.S.) for his work on enzymes in 1926.

Additional reading: Asimov. *Asimov's new guide to science;* p. 572–573.

Perception. *See* Artificial Intelligence.

Percussion
MEDICINE

Leopold Auenbrugger (Austria) developed the method of percussion in 1761. He used percussion as a diagnostic aid in which he tapped portions of the chest and made judgments based on sounds that he heard. Auenbrugger's technique was ignored until 1801 when Jean Nicolas Corvisart (France) revived it. At that time, percussion became widely adopted as a diagnostic tool for chest ailments.

Additional reading: Ackerknecht. *A short history of medicine;* p. 136, 151.

Periodic Table of Elements
CHEMISTRY

In an effort to classify elements, the Russian chemist Dmitri Mendeleev constructed a table in 1869 which arranged the elements in order of atomic weights. His main contribution was to show that elements with similar properties could be arranged to fall into the same columns. He predicted the discovery of elements to fill the gaps; four years later the first of these, **gallium,** was identified. The original skepticism that greeted his discovery gave way to honors and worldwide acclaim.

Additional reading: Feldman. *Scientists & inventors;* p. 186–187.

Personal Computer. *See* Microcomputer.

Pesticide
AGRICULTURE

The first large-scale eradication of a pest by chemical means occurred in Europe in the 1840s when the vine powdery mildew, Unciluna necator, was sprayed with lime sulfur. Chlorinated insecticides were introduced with the discovery of **DDT** in the 1940s. Organic phosphorus insecticides were also introduced in the 1940s. The first to be marketed was a systemic pesticide trade named "Schradan."

Additional reading: Carson. *Silent spring;* p. 16–37.

Petersen Scale
OCEANOGRAPHY

The difficulties of separating estimations of both wind scale and sea scale values led to adoption at the 1939 International Meteorological Conference in Berlin of the Petersen scale, which contained specifications for the visible effects of wind on the sea, allowing for relating the Beaufort and Douglas scales. The Petersen scale had been devised (1939) by the German sailing ship master Captain Petersen. Then in 1947 the World Meteorological Organization adopted a single wave code for observing and recording waves only. (*See also* Beaufort Wind Scale; Douglas Sea Scale)

Additional reading: Fairbridge. *Encyclopedia of oceanography;* p. 786–792.

Petri Dish
MICROBIOLOGY

Julius Richard Petri (Germany) introduced the petri dish in 1887 while working with fellow countryman Robert Koch. Petri's invention served to protect bacteria cultures from being contaminated by airborne bacterial spores.

Additional reading: Asimov. *Asimov's new guide to science;* p. 657.

Petrochemical. *See* Oil Refining.

Petroleum Refining. *See* Oil Refining.

Phage. *See* Bacteriophage.

Phagocyte. *See* Leukocyte.

Pharmacopoeia
PHARMACOLOGY

Greek physician Diocles, who was born around 350 B.C., wrote the first plant identification manual that included information on nutrition and medical usage. Diocles' work constituted the authority on pharmacy until Dioscorides, later Greek physician who lived from around 40 to 80 A.D., wrote De Materia Medica in five volumes. Dioscorides' treatise was the first systematic pharmacopoeia and covered close to 600 plants and 1,000 drugs.

Additional reading: Bettmann. *A pictorial history of medicine;* p. 37.

Phase Rule. *See* Chemical Reaction.

Phenobarbital. *See* Barbiturate.

Phenotype
GENETICS

In 1909 Danish botanist Wilhelm Ludwig Johannsen coined the term "phenotype" to distinguish between the observable characteristics of an organism and its **genotype**, the genetic makeup of an organism.

Additional reading: Dunn. *A short history of genetics;* p. 88–95.

Phlogiston
CHEMISTRY

A false concept of **combustion** was accepted by chemists for a large part of the seventeenth and eighteenth centuries. It was originated by the German scientist Johann Becher, who theorized in 1669 that any material will burn if it contains a substance (soon to be called phlogiston), which escapes as the material burns. The German chemist Georg Stahl gave phlogiston its name around 1687. This concept ignored the fact that metals gain weight when they burn. Phlogiston was not disproved until 1778 when the French chemist Antoine Lavoisier discovered oxygen and showed that burning metals as well as certain other substances combine with oxygen to form new compounds, thus increasing in weight, contrary to the phlogiston theory.

Additional reading: Asimov. *Asimov's biographical encyclopedia;* p. 143, 161–162, 222–226.

Phonograph
ELECTRONICS

In 1877 Thomas Alva Edison (U.S.) patented the phonograph, using tin foil wrapped around cylinders. By 1885 the wax cylinder was invented by Charles Sumner Tainter (U.S.). In 1887 Emile Berliner (U.S.) devised wax-coated disks for use in his device, the gramophone. Another improvement of his, made in 1888, was to record sound so that the grooves on the record involved a lateral side-to-side motion, versus the up-and-down motion used until then. In 1925 electric recording by means of a microphone was invented, then in 1948 the Hungarian-born Peter Goldmark (U.S.) devised the long-playing record. Beginning around 1950 public interest in high fidelity audio equipment began to increase.

Additional reading: Asimov. *Asimov's new guide to science;* p. 427.

Phosphorus
CHEMISTRY

Phosphorus, atomic number 15, can be found in several colors (white, black, and red-violet). It was first identified in 1669 by a German merchant named Hennig Brand. He discovered it while distilling urine with sand and coal. Phosphorus is obtained chiefly from the treatment of phosphate rock, now done in electric furnaces. About 80% of all phosphorus is used in **fertilizers**, with the rest used in fuel additives and industrial compounds. Chemical symbol is P.

Additional reading: *Van Nostrand's scientific encyclopedia;* p. 2204–2209.

Photocopying. *See* Photography; Xerography.

Photoelectric Cell
ELECTRONICS

The knowledge of the **photoelectric effect** led to the invention of the photoelectric cell in 1883 by the American inventor Charles Fritts. It provided a means for controlling an electric current by the intensity of light striking the cell. Selenium, the first light-sensitive material used, was replaced in time by materials even more sensitive. (*See also* Scanning Disk)

Additional reading: Feldman. *Scientists & inventors;* p. 238–239.

Photoelectric Effect
PHYSICS

The process of light causing certain metals to emit **electrons**, called the photoelectric effect, was discovered by the German physicist Philipp Lenard in 1902. The fact that the intensity of the light rather than its wave length was responsible for the energy of the electrons, was explained in 1905 by Albert Einstein (U.S.) by means of the **quantum theory.**

Additional reading: Asimov. *Asimov's new guide to science;* p. 387–389.

Photography
GRAPHIC ARTS

The photographic process became a reality when the French physicist Joseph Niepce made the first photograph around 1826. By 1833 the French physicist Louis Daguerre was able to reproduce images through an elaborate procedure. In 1839 the English scientist William H. F. Talbot invented the negative-positive process, the basis for present-day photography. That year saw the perfecting of the "hypo" (hyposulfite of soda) as a fixing agent in processing, created by the English scientist Sir John Herschel. In 1888 George Eastman (U.S.) developed the Kodak camera for use by amateurs. The development of photocopying machines around 1950 was an offshoot of photography. (*See also* Aerial Photography; Color Photography; Infrared Photography; Instant Camera; Xerography)

Additional reading: Downs. *Landmarks in science;* chap. 53.

Photosynthesis
BIOCHEMISTRY

Dutch physician and plant physiologist Jan Ingenhousz showed in 1779 that green portions of plants give off **oxygen** in the presence of sunlight. In its absence, the roots, flowers, and fruits give off carbon dioxide. In 1837 Henri Dutrochet (France) showed that carbon dioxide is absorbed by only plant cells containing **chlorophyll.** In 1862 German botanist Julius von Sachs established that starch found in green plant cells is a product of photosynthesis. During the 1940s, American biochemist Melvin Calvin discovered and isolated the intermediate products of photosynthesis and described how green plants take carbon dioxide from the air and combine it with water to form starch. He was awarded the 1961 Nobel Prize in chemistry for his work.

Additional reading: Asimov. *Asimov's new guide to science;* p. 588–594.

Phototypesetting
GRAPHIC ARTS

The process of projection onto photographic film of symbols and characters to produce printed materials is called phototypesetting. One of the earliest such devices was patented in 1895 by William Friese-Greene (England). Several models were invented over the years. A major improvement came with the Photon system, invented in 1950 by the French inventors Rene A. Higonuet and Louis Morpond. Newer models have used computer-stored systems in which computers generate the required characters from a great number of discrete parts which control the creation of characters by an electron beam. In 1963 U. S. newspapers began to use computerized systems.

Additional reading: *Illustrated science and invention encyclopedia;* v. 13, p. 1743–1745, v. 22, p. 16–18.

Photovoltaic Cell. *See* Solar Battery.

Pi
MATHEMATICS

The origin of the concept of pi (the ratio of the length of the circumference of a circle to its diameter) is unknown, but the most accurate calculation of it in classical times was made by Archimedes, a Greek mathematician and engineer, whose work was done around 250 BC. Elaborate, lengthy calculations of pi have since been made.

Additional reading: Asimov. *Asimov's biographical encyclopedia;* p. 29–32.

Pi-meson. *See* Meson.

Piezoelectric Effect
SOLID-STATE PHYSICS

Some crystalline materials, when subjected to mechanical pressures, will create an electric current. It was discovered around 1880 by the French physicists and brothers Pierre and Jacques Curie while they were studying crystals that have no center of symmetry. Sound vibrations can also cause such an electrical effect, and this led to the creation of some detectors using this phenomenon. High frequency sound or **ultrasonics** has been used to detect submarines. The opposite effect (using an alternating current to cause a crystal to vibrate) also exists. The crystals used as detectors or generators are called **transducers.** The phenomenon is clearly related to the **electrostrictive effect.**

Additional reading: *Illustrated science and invention encyclopedia;* v. 13, p. 1757–1760; v. 19, p. 2515–2518.

Pion. *See* **Meson.**

Pistil. *See* **Plant Anatomy.**

Pistol. *See* **Revolver.**

Planet. *See* **Planetary Masses; Planetary Motion.**

Planetary Masses — ASTRONOMY

Using his law of gravitation, the English scientist Sir Isaac Newton was able around 1665 to estimate the masses of the planets and of the earth. The work of the English chemist and physicist Henry Cavendish around 1798 led to quite accurate calculations of the mass and density of the earth. (*See also* Gravitation)

Additional reading: Asimov. *Asimov's biographical encyclopedia;* p. 152; 200–202.

Planetary Motion — ASTRONOMY

Centuries of ignorance about the true nature of the motions of the planets were ended when the German astronomer Johannes Kepler discovered in 1609 that planets moved in elliptical orbits. His writings also stated the rules governing their speed of motion. This led to the complete acceptance of the 1530 theory of the Polish astronomer Nicolaus Copernicus regarding the **solar system.** The French astronomer and mathematician Pierre S. Laplace did much to explain perturbations in movements of the planets in 1787. He was working at about the same time as the French astronomer and mathematician Joseph L. Lagrange, whose findings were similar to his. (*See also* Motion)

Additional reading: Asimov. *Asimov's biographical encyclopedia;* p. 105–108; 209–211; 234–236.

Plant Anatomy — BOTANY

German botanist Hieronymus Tragus advanced the study of plant anatomy during the 1500s when he discovered the stamen and pistil as definable organs. In 1682 English botanist Nehemiah Grew published *The Anatomy of Plants,* which described sexual reproduction in plants. Grew established that the stamen and pistil were male and female organs in plants. Rudolph Jakob Camerarius (Germany) supported the existence of sexuality in plants in his 1694 volume entitled De sexu plantarum. He also reported on early pollination experiments.

Additional reading: Greene. *Landmarks of botanical history;* p. 304–359, 973–974.

Plant Classification. *See* **Animal and Plant Classification.**

Plant Cytology — CYTOLOGY

German botanist Eduard Adolf Strasburger was said to have launched plant cytology as a discipline with his 1875 publication Zellbildung und Zelltheilung. Strasburger described cell division in the embryo of a conifer, and illustrated a fibrous spindle at several stages of **mitosis.**

Additional reading: Hughes. *A history of cytology;* p. 62–63.

Plant Nutrition — AGRICULTURE

In 1886 German chemist Hermann Hellriegel reported his discovery that certain legumes, such as peas and beans, used atmospheric nitrogen as a source of nutrition. Hellriegel established that the legumes would tend to refertilize the soil with the nitrogen, lessening the need for chemical fertilizers. (*See also* Fertilizer)

Additional reading: Morton. *History of botanical science;* p. 420.

Plant Physiology — BOTANY

Plant physiology was founded by English botanist and chemist Stephen Hales during the 1700s. Hales was the first to recognize that a portion of the air contributed to the nourishment of plants.

Additional reading: Hawks. *Pioneers of plant study;* p. 227–231.

Plasma — PLASMA PHYSICS

Plasma is the name for a gas at extremely high temperatures (millions of degrees) which has all its **electrons** stripped from its atoms. The term was proposed around 1930 by the American physicist Irving Langmuir. It is currently studied in controlled **nuclear fusion,** and is sometimes called the fourth state of matter.

Additional reading: Asimov. *Asimov's new guide to science;* p. 502–503.

Plaster of Paris — MEDICINE

Between 845 and 930 A.D. Rhazes (Persia) prepared plaster of Paris to form casts used to hold broken bones in place. Flemish army surgeon Anthonius Mathijsen introduced a quick-setting plaster of Paris in 1852 used as a substitute for braces and splints to immobilize injured bones.

Additional reading: Bick. *Source book of orthopaedics;* p. 286–290.

Plate Glass — MATERIALS

Large sheets of glass had to be made by glass blowing until around 1700, when glass makers in France developed a system of pouring molten glass onto a casting table, which had movable guides for determining the sheet size. In 1773 an English company began using this technique, making sheets up to 15 feet long. By 1789 a steam engine developed by Matthew Boulton (England) and James Watt (Scotland) was used to grind and polish the sheets. Blown glass was used for windows in Great Britain until 1845, when a window tax was repealed, leading to use of cast glass. The scientific study of the composition of glass was begun in the 1860s, and in 1875 a formula for window glass was developed and has been used more or less ever since.

Additional reading: *Illustrated science and invention encyclopedia;* v. 9, p. 1122–1126.

Plate Tectonics — GEOLOGY

Geologists now generally accept the concept called "plate tectonics" which affirms that the earth is divided into six or so rigid sectors which are moving with respect to each other. They are separated by the mid-ocean rift, major mountain chains and shear zones. The continents can be compared to huge icebergs carried along by the currents underneath them. The term "plate tectonics" was first used by D. P. McKenzie (England) and R. L. Parker (U.S.) in a journal article in 1967. Support for this concept comes from many sources, including studies of the movement of continents as shown in **magnetic reversal.** It also helps explain the phenomenon of **continental drift** as well as **earthquakes.**

Additional reading: Gribbin. *Our changing planet;* p. 59, 67–74.

Plating. *See* **Electroplating.**

Platinum — CHEMISTRY

A silvery-white, ductile, soft metal, platinum, atomic number 78, was described around 1557 by the Italian physician Julius Caesar Scaliger, but the first substantial experiments with platinum were done in 1750 by the English physician William Brownrigg. Platinum is a whitish steel-gray metal, usually found alloyed with other elements, particularly metals. It is used in high-octane gasoline, in vitamins, in jewelry, in devices needing high melting points, and in electrical equipment. Chemical symbol is Pt.

Additional reading: *Van Nostrand's scientific encyclopedia;* p. 2258–2260.

Plow
AGRICULTURE

The earliest plows appear to have been built between 4000 and 3000 B.C. The "spade plow" originated in Mesopotamia and Egypt and was an adaptation of the digging stick. The "crook plow" developed about the same time and place and was derived from a forked branch that was tied to a pole. The junction of the forks acted as a **plowshare** and dug into the soil. The first plows were probably dragged by humans who were soon replaced by pairs of oxen. During the Middle Ages the Slavs invented the "mold-board plow" which consisted of a twisted plank resting behind a plowshare. The plank could force the loosened earth up and over as the plow passed.

Additional reading: Curwen. *Plough and pasture;* p. 61–92.

Plowshare
AGRICULTURE

The iron plowshare appears to have originated in Britain between 400 and 1 B.C. It was 12 to 18 inches in length and 2 inches wide at the socket. It may have been mounted on a crook plow. In 1785 the English inventor Robert Ransome patented a cast-iron plowshare, while in 1803 he also patented a self-sharpening plowshare. (*See also* Plow)

Additional reading: Curwen. *Plough and pasture;* p. 61–92.

Pluto
ASTRONOMY

The discrepancies in the orbit of Uranus, once thought to be caused by Neptune alone, were believed by the American astronomer Percival Lowell to be also caused by still another planet beyond Neptune. Lowell searched for years before his death in 1916, but not until 14 years later did American astronomer Clyde Tombaugh discover Pluto (in 1930). The Dutch-born astronomer Gerard Kuiper (U.S.) determined in 1950 that Pluto had a diameter of only 3,600 miles. It was so small it could not have appeciably affected the orbit of Uranus. Then in 1978 the American astronomer James Christy discovered that Pluto had a satellite, the only one discovered to date.

Additional reading: Asimov. *Asimov's new guide to science;* p. 144–146.

Plutonism
GEOLOGY

Plutonism is the theory once held by some geologists that all the earth's rocks were formed by the earth's internal heat acting through volcanic vents. Thus the rocks were all supposed to have solidified from a molten mass. One of the earliest proponents was Anton-Lazzaro Moro (Italy), who wrote on the subject in 1740. Some credit the origin of the theory to the Scottish geologist James Hutton, writing in 1785. Hutton's theories were attacked by his peers, such as the Irish mineralogist Richard Kirwan around 1793. In time plutonism was abandoned by geologists.

Additional reading: Faul. *It began with a stone;* p. 63, 108–110.

Plutonium
CHEMISTRY

Plutonium, a man-made radioactive element, atomic number 94, was discovered in 1941, by the American chemists A. C. Wahl, G. T. Seaborg, and J. W. Kennedy, and perhaps others. It can be fissioned by both high and low energy neutrons, thus making it valuable as a nuclear fuel. However, it is a very toxic substance. Most of it is derived from the processing of **uranium** used as fuel at **nuclear reactors**. Chemical symbol is Pu.

Additional reading: *Van Nostrand's scientific encyclopedia;* p. 2262–2264.

Polarized Light
OPTICS

Light rays vibrate in every direction, but certain materials can absorb all the vibrations except those in one plane. The resulting light is said to be polarized. The explanation of polarized light had come from the French physicist Augustin Fresnel around 1814 when he showed that light vibrated at right angles to the line of propagation and some planes of vibration can be absorbed. An early device demonstrating this was the Nicol prism invented by the Scottish physicist William Nicol in 1829. In 1815 the French physicist Jean Baptiste Biot showed that some substances twist the plane of polarization. Manmade materials which polarized light were invented in 1932 by the U. S. physicist Edwin Land. He developed products like camera filters and sunglasses which used such materials.

Additional reading: Asimov. *Asimov's new guide to science;* p. 512–514.

Polaroid Process. *See* Instant Camera.

Polio Vaccine
IMMUNOLOGY

In 1952 an American microbiologist, Jonas Edward Salk, prepared the first vaccine against poliomyelitis. After trying it with children who had already survived polio and who would therefore be resistant, he inoculated children without a history of the disease. In 1954 the vaccine was produced in large quantities, and by 1955, it was being widely used. Salk's vaccine used a dead virus as the immunization agent. In 1957 Albert Sabin, a Polish-born American, developed a vaccine that used a live virus and was administered orally. In 1960, after trying the oral polio vaccine in eastern Europe, it was introduced to the United States. (*See also* Virus Culture)

Additional reading: Robinson. *The miracle finders;* p. 39–47.

Pollen
BOTANY

In the 1500s, German botanist Valerius Cordus used the name "pollen" to describe what he had observed as flower dust. Pollen turned out to be the male reproductive bodies produced in a seed plant's pollen sacs.

Additional reading: Greene. *Landmarks of botanical history;* p. 390–392.

Pollination. *See* Plant Anatomy.

Pollution Control
ENGINEERING

For centuries civilization has endured the pollution of air and water. Smoke pollution was a concern as far back as 1661 when John Evelyn (England) proposed methods of smoke prevention. In 1715 a smoke prevention damper for use on chimneys was invented in France. Various regulations have been created since then around the world. Automobile emission controls were developed in the United States in the 1960s. As far back as the 1800s concern over water pollution resulted in such laws as the act of 1856 to control pollution of the Thames River in London. Removal of impurities in water supplies by filtering took place in London in 1828. Aeration and chemical treatment followed in later years. (*See also* Waste and Sewage Disposal)

Additional reading: *Illustrated science and invention encyclopedia;* v. 14, p. 1809–1815; v. 19, p. 2616–2620.

Polonium
CHEMISTRY

A radioactive element, polonium, atomic number 84, was discovered by the French chemists Pierre and Marie Curie in 1898 because of its **radioactivity**. They extracted it as an ingredient of pitchblende, but it occurs in nature only as a decay product of **thorium** and **uranium**. It is used chiefly as a laboratory source of **neutrons** and in meteorological instruments. Chemical symbol is Po.

Additional reading: *Van Nostrand's scientific encyclopedia;* p. 2281.

Polygraph
PSYCHOLOGY

John Augustus Larsen (U.S.) and Leonard Keeler (U.S.) invented the first polygraph in 1921 to detect changes in blood pressure, pulse rate, breathing rate, and perspiration brought on by emotions. Its usual application is to detect emotions evoked by telling a lie.

Additional reading: Asimov. *Asimov's new guide to science;* p. 833–836.

Porcelain
MATERIALS

Porcelain, generally considered the most desirable form of pottery, is often called chinaware because it originated in China. There are three kinds of porcelain. Hard-paste porcelain was made in China, starting probably around 900 A.D. Kaolin and China stone were used to produce translucent, glasslike china. It became popular in Europe after 1600. In 1710 Johann Boettger (Germany) developed a good imitation, called Dresden or Meissen china, using marble as a fluxing agent. The second type is soft-paste porcelain, made from substitutes for the Chinese materials. One source was Sevres, France, then it began to be made in England around 1700. The third type is called bone china, using bone ash. It was created in 1750 and then perfected by Josiah Spode (England) in 1799.

Additional reading: Derry. *Short history of technology;* p. 93, 583–592.

Portland Cement. *See* Concrete and Cement.

Positron
PARTICLE PHYSICS

The efforts of physicists to understand **cosmic rays** led accidentally to the discovery of a new particle, the positron. In working with a **cloud chamber** the American physicist Carl D. Anderson in 1932 arranged for cosmic rays to bombard a lead plate and then enter the chamber. In the bombardment they knocked particles out of the lead plate, a few of which particles curved in the wrong direction for a negatively charged particle and thus could not be an **electron.** He concluded he had found a positively charged opposite to the electron. Later named the positron, Anderson's particle confirmed a prediction made in 1930 by the English physicist Paul Dirac about the existence of **antiparticles.**

Additional reading: Asimov. *Asimov's new guide to science;* p. 331–334.

Postmortem
MEDICINE

A physician from Cremona, Italy, performed the first postmortem examination in order to discover the cause of an epidemic spreading through the city. French physician Jean Francois Fernel was the first modern physician to make postmortems a regular procedure in his clinical practice. He reported his findings in a book published in 1542. Italian anatomist Giovanni Battista Morgagni published *The Seats and Causes of Diseases Investigated by Anatomy* in 1761 and based his findings on 640 postmortem examinations he had performed. Morgagni asserted that diseases are situated in certain organs of the human body. (*See also* Dissection)

Additional reading: McGrew. *Encyclopedia of medical history;* p. 68–69, 236–241.

Potassium
CHEMISTRY

Potassium, a silver-white metal, atomic number 19, oxidizes instantly on exposure to air and reacts violently with water. It was discovered in 1807 by the English chemist Sir Humphry Davy. Potassium does not occur in nature in the free state due to its exceptional chemical reactivity. It is prepared by the treatment of minerals such as potash (potassium carbonate) and other deposits of potassium salts, found in some western states and in the Dead Sea. It is used in plasma-based power generation and in heat exchange media. Chemical symbol is K.

Additional reading: *Van Nostrand's scientific encyclopedia;* p. 2295–2297.

Potter's Wheel
MECHANICAL ENGINEERING

The reliance of civilizations on pottery for cooking and containment of liquids dates back to the earliest times. Therefore it should not be surprising that the potter's wheel for shaping the product was developed as long ago as 3000 B.C. in Egypt. The earliest ones were spun by hand, but by 2000 B.C. a second wheel was introduced, placed under the first and probably turned by the potter's foot or by an assistant. After the fall of the Roman Empire, the potter's wheel disappeared except in some scattered parts of Europe, where the kick-wheel (foot-operated) left both of the potter's hands free. Not until the ninth century did the wheel return to Great Britain.

Additional reading: Derry. *Short history of technology;* p. 75–76; 91–92.

Praesodymium
CHEMISTRY

Colored silver-gray, the metallic praesodymium, atomic number 59, must be handled in inert atmospheres or in a vacuum. It was discovered in 1885 by the Austrian chemist Baron Auer von Welsbach. It is obtained by treatment of the minerals bastnasite and monazite. Its uses include carbons for arc lights, optical glass polishing compounds, petroleum refining, and permanent magnet compounds. Chemical symbol is Pr.

Additional reading: *Van Nostrand's scientific encyclopedia;* p. 2313.

Prednisone. *See* Kidney Transplant.

Primary Production
ECOLOGY

American limnologists E. A. Birge and Chancey Juday developed the concept of primary production during the 1920s while measuring energy budgets in lakes. They defined primary production as the rate at which food energy is generated by **photosynthesis.**

Additional reading: McIntosh. *The background of ecology;* p. 96–98.

Printing
GRAPHIC ARTS

One of the world's most important inventions was the development of a printing process using movable type, thus enabling printers to produce books much more quickly than by using the laborious block printing method. Both speed and uniformity of work were improved. Although authorities differ, the originator is generally agreed to be the German printer Johannes Gutenberg, whose Bible, produced around 1450, won him international fame. Many developments in **printing presses** followed the Gutenberg era. (*See also* Computer Printer; Lithography; Typesetting)

Additional reading: Feldman. *Scientists & inventors;* p. 16–17.

Printing Press
GRAPHIC ARTS

For nearly 300 years, printing was done laboriously on hand-operated presses. In 1814 the German engineer Friedrich Koenig invented the first steam-powered printing press. By 1824 the British inventor William Church devised a system for mechanizing the pick-up of paper. A major change came about in 1846 with the development by the American inventor Richard Hoe of the rotary press, replacing the flat bed system, resulting in much higher speeds. In 1865 another American inventor, William Bullock, devised a way for the press to use a continuous roll of printing paper. Numerous improvements have been made since then.

Additional reading: Feldman. *Scientists & inventors;* p. 104–105.

Procaine Hydrochloride. *See* Novocain.

Promethium
CHEMISTRY

A soft, silver-white metal, promethium, atomic number 61, was discovered in 1945 by the U. S. physicists J. A. Marinsky and L. E. Glendenin by bombardment of uranium. The naturally occurring **isotope** is radioactive and consequently needs to be handled in a shielded area. The most common sources are **uranium** or **plutonium** wastes from which it must be extracted. One of its uses is for power sources for surgical implants. Chemical symbol is Pm.

Additional reading: *Van Nostrand's scientific encyclopedia;* p. 2331.

Prontosil. *See* Sulfanilamide.

Propeller
MECHANICAL ENGINEERING

Marine propellers have been used for ships since about 1838 when the British ship *Archimedes* became the first propeller-driven steamship. However, early experiments date back to Archimedes (Greece), who demonstrated around 230 B.C. that a helical blade could move water. In 1768 the French engineer Alexis Jean Pierre Paveton proposed using propellers for ship propulsion, and in 1796 John Fitch (U.S.) experimented with propellers. The use of propellers for **aircraft** dates back to the first powered flights by the

Wright brothers in 1903, although a propeller-driven steam-powered aircraft was designed in 1842 by W. S. Henson (England); it was never built. **Jet engines** have today greatly diminished the use of propellers for aircraft.

Additional reading: *Illustrated science and invention encyclopedia;* v. 1, p. 40–44; v. 11, p. 1454–1456; v. 14, p. 1870–1874.

Prostaglandins

BIOCHEMISTRY

Prostaglandins, compounds that affect the human nervous system, circulation, female reproductive organs, and metabolism were discovered in the 1930s by American physicians Ralph Kurzrok and C.C. Lieb in human semen. In 1935 Swiss physiologist Ulf von Euler found it in sheep semen and named it prostaglandin. Another Swiss scientist, Sune Bergstrom, discovered in 1949 that prostaglandins were a group of compounds, and isolated one of those compounds in pure crystal form in 1957. In 1959 Bergstrom had deciphered its molecular structure. In 1966 Ugandan pharmacologist Sultan M.M. Karim discovered that prostaglandins were connected to the commencement of labor and to spontaneous abortions.

Additional reading: Robinson. *The miracle finders;* p. 281–286.

Protactinium

CHEMISTRY

Protactinium, atomic number 91, is a radioactive metal of the actinide series. In 1917 it was discovered independently and almost simultaneously by the German chemist Otto Hahn with Austrian physicist Lise Meitner, by the American chemist Kasimir Fajans, and by the English chemist Frederick Soddy with the Scottish chemists John A. Cranston and Alexander Fleck. It had been predicted as early as 1871 by Dmitri Mendeleev on the basis of his periodic table. It is found in nature in all **uranium** ores, from which it must be extracted. It is a dangerously toxic substance. Although no known uses are reported, in recent years it has been discovered to be a superconductor under certain conditions. Chemical symbol is Pa.

Additional reading: *Van Nostrand's scientific encyclopedia;* p. 2334–2335.

Protein

BIOCHEMISTRY

In 1838 a Dutch chemist, Gerardus Johannes Mulder, introduced the word "protein" to describe what he took to be the basic building block of the albuminous substances he was studying. While Mulder was mistaken in his theory of the structure of the albuminous substances, the word "protein" was eventually used for the substances themselves. In 1907 Emil Hermann Fischer, a German chemist, built a protein molecule made up of 18 amino acid units to demonstrate the way in which one **amino acid** combines with another. Fischer's model initiated the field of protein structure.

Additional reading: Asimov. *Asimov's new guide to science;* p. 551–554.

Proton

NUCLEAR PHYSICS

A long study of **alpha particles** was made by the British physicist Ernest Rutherford who stated around 1914 that they were helium atoms with their **electrons** removed. Since these positive-charged particles behaved like the positive rays discovered by the German physicist Eugen Goldstein in 1886, Rutherford declared in 1914 that the positive rays were a new particle which he called a proton, two of which were found in the helium nucleus. In 1955 scientists discovered the **antiproton.**

Additional reading: Asimov. *Asimov's new guide to science;* p. 318–319.

Protoplasm

CYTOLOGY

In 1839 the Czech physiologist Jan Evangelista Purkinje gave the name "protoplasm" to the colloidal fluid found in certain cells. In 1846 Hugo von Mohl, a German botanist, extended its use to signify the contents of all cells, including those of plants. Thomas Henry Huxley, an English botanist, introduced the term to the general public in 1868 at Edinburgh in a talk entitled "The Physical Basis of Life."

Additional reading: Magner. *A history of the life sciences;* p. 214–236.

Protozoa

ZOOLOGY

Anton van Leeuwenhoek, a Dutch biologist and microscopist, announced the discovery of one-celled creatures he called "animalcules" in 1676. One year later, in 1677, the English physicist Robert Hooke demonstrated van Leeuwenhoek's discovery of protozoa to the Royal Society of London. The first time in which a protozoon rather than a bacterium was shown to cause disease was in 1880 when Charles Louis Alphonse, a French physician, found it to be a causative factor in malaria.

Additional reading: Feldman. *Scientists & inventors;* p. 44–45.

Proust's Law. *See* Constant Composition (Chemical).

Psychiatric Asylum

MEDICINE

The first known mental hospital was opened in 1409 in Valencia, Spain, by Fray Juan Galiberto Jofre, a monk of the Order of Our Lady of Mercy. The hospital itself was called Our Lady of the Innocents. Jofre was moved to establish the asylum after seeing a group of children harrass a mentally disturbed person.

Additional reading: *World history of psychiatry;* p. 97–118.

Psychiatry

PSYCHIATRY

The earliest glimmerings of psychiatry can be traced to around 1950 B.C. in the Code of Hammurabi, the King of Babylonia. The Code recommended that **opium** and olive oil be used to treat psychiatric disorders which, at that time, were blamed on demonic possession. During the 1870s and 1880s Jean Martin Charcot, a French physician, introduced modern psychiatry as an area of medicine that dealt with mental disorders. Charcot worked with hysterics.

Additional reading: Murray. *A history of Western psychology;* p. 287–293.

Psychoanalysis

PSYCHOLOGY

Psychoanalysis, a technique used to treat neuroses and other emotional disorders, began in the 1890s when Sigmund Freud (Austria) learned about the use of posthypnotic suggestion from Jean Martin Charcot (France) and about talking cures and **catharsis** from Josef Breuer (Austria). In 1895 Freud published *Studies in Hysteria* and therein introduced psychoanalytic techniques involving the **unconscious.** (*See also* Hypnosis)

Additional reading: Murray. *A history of Western psychology;* p. 287–314.

Psychology

PSYCHOLOGY

Marko Marulic, a poet from Dalmatia, is given credit for being the first to use the word "psychology" in his Latin treatise entitled *Psichiologia de ratione animae,* published around 1520. In 1905 William McDougall (England) defined psychology as the study of behavior.

Additional reading: Murray. *A history of Western psychology;* p. 8–86.

Psychology Laboratory

PSYCHOLOGY

In 1879 German psychologist Wilhelm Max Wundt established the first laboratory to be devoted to experimental psychology. It was situated at Leipzig University. American psychologist Granville Stanley Hall knew of Wundt's laboratory and established the first one in the United States at Johns Hopkins University in 1883.

Additional reading: Murray. *A history of Western psychology;* p. 171–173, 193.

Psychophysical Method
PSYCHOLOGY

German physicist Gustav Theodor Fechner introduced psychophysical methods in order to study the relationship between external stimulus and subjectively experienced optical, acoustical, or tactile sensation. Fechner's psychophysical methods were significant because they introduced experimental methodology to the study of psychology. Fechner reported on his research in a book published in 1860.

Additional reading: Alexander. *The history of psychiatry;* p. 166–167.

Psychosurgery. *See* **Lobotomy.**

Pulley
MECHANICAL ENGINEERING

The use of pulleys dates back to ancient times. Some historians credit Archimedes (Greece) with having invented them around 225 B.C. Probably made of wood, they were used on early cranes, such as those mentioned in writings by Vitruvius (Italy) around 30 B.C. However, the works of Homer indicated that they were used on Greek ships to hoist sails as far back as 600 B.C.

Additional reading: *Illustrated science and invention encyclopedia;* v. 14, p. 1880–1881.

Pulsar. *See* **Neutron Star.**

Pulse Watch
MEDICINE

Nicholas Cusa (Germany) recommended counting pulse rate as a diagnostic aid around 1450. In 1603 Santorio Santorio, also known as Sanctorius, an Italian physician, described and invented the "pulsilogium" for counting pulse beats. His instrument was inspired by the pendulum of Galileo (Italy) and was used specifically by him to gather data allowing the comparison of pulse beat with disease. The use of the pulsilogium languished until 1707 when John Floyer (England) began counting pulse beats with the aid of "pulse watches." (*See also* Sphygmometer)

Additional reading: Bettmann. *A pictorial history of medicine;* p. 140, 143.

Pulsejet. *See* **Jet Engine.**

Pump. *See* **Steam Pump; Water Pump.**

Punched Card
COMPUTER SCIENCE

Punched cards are usually associated with computers, but they have a long history, dating back at least to 1725, when the French inventor Basile Bouchon used punched paper to control a loom using a particular design. In 1805 the French inventor Joseph Jacquard invented the **Jacquard loom** that was controlled by punched cards. Around 1829 the English mathematician Charles Babbage saw an example of Jacquard's work and sensed the value of cards for a mechanical calculator he was developing. Although he never completed his machine, his plans for cards may have influenced the American statistician Herman Hollerith, who devised a **calculating machine** in 1884 that used punched cards. Card use for computer input began declining in the 1970s.

Additional reading: *Illustrated science and invention encyclopedia;* v. 4, p. 481–484.

Purine. *See* **Nucleic Acid.**

Pyramid
CIVIL ENGINEERING

These impressive structures, designed to be safe tombs for the remains of Egyptian pharaohs, represent excellent examples of the ability of ancient engineers to overcome tremendous difficulties. Little is known of how heavy materials were transported and prepared. Imhotep is believed to have been the builder of the first pyramid, constructed in Egypt around 2700 B.C. It was 200 feet high with a base of 358 by 411 feet. The most famous example is the Great Pyramid of Gizeh, which has a height of about 463 feet. Certain tribes in America also built pyramids.

Additional reading: Clark. *Works of man;* p. 15–18.

Pyrilamine. *See* **Antihistamine.**

Pyrimidine. *See* **Nucleic Acid.**

Q

Quality Control
INDUSTRIAL ENGINEERING

Quality control refers to the techniques and processes used by manufacturers to ensure consistent high standards for their products. One of the earliest examples was a program cosponsored in the United States in 1916 by the Western Electric Company and Bell Telephone Laboratories. In 1924 Walter Shewhart (U.S.) developed statistical control charts that aided sampling of large outputs, a process further aided in 1973 by the theory of Abraham Wald (U.S.) on scientific sampling. Around 1944 methods of this sort previously developed by H. F. Dodge and H. G. Romig (both U.S.) came into prominence, particularly in World War II production projects. Reliability, the ability of a product to function consistently over its life span, also received attention about this time.

Additional reading: *Illustrated science and invention encyclopedia;* v. 14, p. 1893–1894.

Quantum Mechanics. *See* **Wave Mechanics.**

Quantum Theory
PHYSICS

The long controversy over whether light was a type of wave or was made up of particles was partially answered by the work of the German physicist Max Planck in 1900. Planck proposed that radiation, including light, was emitted in small units that he called quanta. The wave length of the radiation could determine the amount of energy in a quantum. His theory later helped explain the behavior of atoms and nuclei and was verified by the studies of Albert Einstein (U.S.) on the **photoelectric effect** in 1905. In 1913 the Danish physicist Niels Bohr applied quantum theory in explaining the structure of atoms, stating that when **electrons** change their orbits, only fixed amounts of energy would be involved, thus supporting Planck's theory of quanta.

Additional reading: Asimov. *Asimov's new guide to science;* p. 386–389.

Quark
NUCLEAR PHYSICS

In the 1950s physicists found the ever-growing number of **subatomic particles** very confusing, so that serious study was given to creating a sort of periodic table for these particles. In 1961 one such table was devised by the American physicist Murray Gell-Mann. Also working at the same time was the Israeli physicist Yuval Ne'emen. They put groups of particles together, forming a symmetrical pattern. To account for all places in the table, Gell-Mann predicted in 1964 the existence of the needed particles. He gave them the whimsical name of quarks, three of them needed to explain the existence of **neutrons, protons**, and strange particles. Subsequently a great deal of research has been carried out in the search for quarks. A fourth type of quark has subsequently been predicted.

Additional reading: Asimov. *Asimov's new guide to science;* p. 361–165.

Quasar
ASTRONOMY

In 1960 radio astronomers began to note some signals they believed to be coming from individual stars rather than from a galaxy. That same year the American astronomer Allan Sandage found that the spectrum of one of the sources was rich in ultraviolet light. These radio waves seemed not to be ordinary stars, so they were named "quasi-stellar radio sources" which, upon the suggestion in 1964 of the Chinese-born physicist Hong Yee Chiu (U.S.), was shortened to quasar.

Additional reading: Asimov. *Asimov's new guide to science;* p. 66–69.

Quern
AGRICULTURE

The quern, a handmill used to grind grains, appeared between 6000 and 5000 B.C. in Persia and consisted of two stones configured in the shape of a saddle, making a "saddle-quern." The "rotary quern," a revolving handmill, appeared in the Mediterranean region between 5000 and 4000 B.C.

Additional reading: Curwen. *Plough and pasture;* p. 125–134.

Quinacrine
PHARMACOLOGY

German researchers Fritz Mietzsch, H. Mauss, and Walter Kikuth synthesized quinacrine, a new antimalarial drug, with Kikuth publishing the first reports about it in 1932. It was originally marketed as Plasmoquine E and later as Atabrine. Over the years it was used to treat tapeworm infections and amebiasis more than malaria.

Additional reading: Sneader. *Drug discovery;* p. 268–275.

Quinine
PHARMACOLOGY

Thomas Sydenham, an English physician, popularized the use of quinine for treating malaria in the mid-1600s. Quinine, an alkaloid compound extracted from the cinchona tree, was isolated in 1820 by two French chemists, Pierre Joseph Pelletier and Joseph Bienaime Caventou. In 1944 two American chemists, Robert Burns Woodward and William von Eggers Doering, successfully synthesized quinine.

Additional reading: Woglom. *Discoverers for medicine;* p. 132–144.

R

Rabies Vaccine
IMMUNOLOGY

French chemist Louis Pasteur developed a rabies vaccine as an immunization against the fatal disease and used it for the first time on a human in 1885. A nine-year-old French boy, Joseph Meister, had been bitten 14 times by a rabid dog. After Pasteur injected him several times with the pulverized spinal cord of a rabbit that had died of rabies, the boy survived.

Additional reading: Magner. *A history of the life sciences;* p. 261–263.

Rack Railway. *See* **Cog Railway.**

Radar
ELECTRONICS

Although some early experimenters in radio like Marconi noticed in 1922 the effect of reflection of radio waves by metallic objects, it was not until the 1930s that military uses of this phenomenon were studied. The leading investigator was the Scottish engineer Sir Robert Watson-Watt who developed a warning system for enemy aircraft in 1935 that would detect a plane 40 miles away. By 1938 a whole network of stations was in place. The key ingredient for later operation was the development in 1939 of the multicavity magnetron, which had been invented by the English scientists Harry Boot and John Randall. German use of radar was given a low priority, enabling Allied forces to turn back air and rocket attacks. Radar is an acronym standing for RAdio Detection And Ranging.

Additional reading: Feldman. *Scientists & inventors;* p. 284–285.

Radiation. *See* **Black Body Radiation; Radioactivity; Names of specific types of radiation.**

Radio
COMMUNICATIONS

The mathematical treatment of **electromagnetism** of James Clerk Maxwell (Scotland) around 1864 led the German physicist Heinrich Rudolf Hertz to experiment in 1888 with oscillating electric charges in order to emit **electromagnetic waves.** In 1894 the British scientist Sir Oliver Lodge sent Morse-code signals a half-mile. In 1895 Aleksandr Popov (Russia) built a radio, demonstrating it the next year. Perhaps preceding his work was that of Guglielmo Marconi (Italy) who was able in 1895 to transmit a signal beyond the horizon, then in 1901 his signals were sent across the Atlantic. In 1929 Andre G. Clavier (France) began experiments with microwaves. (*See also* Amplitude Modulation; Crystal Rectifier; Electron Tube; Frequency Modulation)

Additional reading: Feldman. *Scientists & inventors;* p. 228–229, 262–263.

Radio Astronomy
ASTRONOMY

Radio astronomy originated at the Bell Telephone Laboratories in 1931 when a radio engineer, Karl Jansky (U.S.), noted a faint steady signal that did not appear to be coming from any usual source. By 1933 he decided the signals must be coming from the Milky Way, particularly from Sagittarius. In 1938 a radio ham, Grote Reber (U.S.), using a homemade parabolic dish antenna, found many sources of the signals, particularly in Cassiopeia. In 1947 the Australian astronomer John Bolton determined that the Crab Nebula was the third strongest source. Further study of radio sources led to the discovery of **quasars.**

Additional reading: Asimov. *Asimov's new guide to science;* p. 61–64.

Radio Tube. *See* **Electron Tube.**

Radioactivity
NUCLEAR PHYSICS

The discovery of **X-rays** in 1895 was due to the fluorescence they caused, and it was this study of fluorescence which caused the French physicist Antoine Becquerel to discover radioactivity in 1896. He found that a **uranium** compound in his laboratory fogged a photographic plate. In 1899 he proved the fogging was caused by particles, and in 1900 he showed that the negatively charged particles were identical to **cathode rays.** The process was named radioactivity in 1898, the name having been proposed by the French physicist Marie Curie. (*See also* Artificial Radioactivity)

Additional reading: Asimov. *Asimov's biographical encyclopedia;* p. 274–276.

Radioimmunoassay
IMMUNOLOGY

During the 1950s American biophysicists Rosalyn Sussman Yalow and Solomon A. Berson developed the technique of radioimmunoassay to measure plasma insulin concentrations in diabetics and normal persons. The technique led to the discovery of **hormones** in previously unknown places in the body. Yalow shared in the 1977 Nobel Prize for medicine and physiology for her work in this area.

Additional reading: Crapo. *Hormones: the messengers of life;* p. 13–14.

Radioisotope
NUCLEAR PHYSICS

The work of the French chemists Frederic and Irene Joliot-Curie on **radioactivity** revealed in 1934 that they had created radioactive isotopes. These artifically produced substances as well as radioactive isotopes that occur naturally in nature are called radioisotopes, of which there are about 50. More than a thousand artifical radioisotopes have been made.

Additional reading: Asimov. *Asimov's biographical encyclopedia;* p. 771–772.

Radiometer
PHYSICS

A radiometer is a device which consists of several pieces of flat metal, or vanes, pivoted to turn, mounted in a vacuum. One side of each vane was blackened so as to absorb heat and the other shiny so as to reflect it. The device worked only when it was exposed to sunlight or other radiation, but it ceased if the vacuum was very good, thus proving it needed a few molecules of air to knock the vanes forward. It provided more evidence of the **kinetic theory of gases.** It was invented in 1875 by the English physicist Sir William Crookes.

Additional reading: Asimov. *Asimov's biographical encyclopedia;* p. 457–458.

Radium
CHEMISTRY

A white metal, radium, atomic number 88, is intensely radioactive; all its isotopes are radioactive. It was discovered in 1898 by the French chemists Pierre and Marie Curie, assisted by M. G. Bemont. It is found in all **uranium** ores, treated usually by addition of a barium salt. Radium is a dangerous material and needs careful handling. It has been used in luminous paints, although less hazardous substitutes have now been found. Chemical symbol is Ra.

Additional reading: *Van Nostrand's scientific encyclopedia;* p. 2400–2401.

Radon
CHEMISTRY

Radon, atomic number 86, is the heaviest of the noble gases. All its **isotopes** are radioactive. It is a colorless gas normally, but when condensed it becomes a fluorescent liquid. It was discovered in 1900 by the German physicist Friedrich Dorn, who noted that radium emanated a gaseous daughter, which was then called "Radium Emanation." In 1910 the Scottish chemist Sir William Ramsay and Robert Whytlaw-Gray, an English chemist, determined the density of the gas. Beginning in 1923 it became known as radon. The natural isotopes are found in soils, the oceans, lakes, and the atmosphere, although the amounts found vary greatly. It has uses in medicine and as a tracer in leak detection, as well as a source of gamma rays in inspecting metal welds. Chemical symbol is Rn.

Additional reading: *Van Nostrand's scientific encyclopedia;* p. 2401.

Railroad Car Brake. *See* **Brake.**

Railroad Passenger Car
MECHANICAL ENGINEERING

From the earliest days of the railroads (1758 in Great Britain) until the end of the nineteenth century most railroad cars were built of wood, but by 1900 steel coaches were common. One American coach at that time weighed more than 80 tons. New lightweight metals and alloys began to be used. In the 1950s the Budd Company (U.S.) built one weighing only 27 tons. About that time the Spanish Talgo cars had an axle at one end with the other end pivoted on the car ahead; in the United States two-level cars with plastic-covered observation domes were created. Sleeping cars were first used in 1836. George Pullman (U.S.) in 1859 first built what became a famous brand of cars. In 1876 the International Sleeping Car Company was formed by the Belgian engineer Georges Nagelmachers.

Additional reading: *Illustrated science and invention encyclopedia;* v. 14, p. 1925–1934.

Railroad Refrigerator Car
AGRICULTURE

J. B. Sutherland (U.S.) patented a railroad car equipped with insulation, ventilation, air circulation, and ice storage in 1867. This represented the first application of refrigeration to railroad cars and allowed for the transport of perishable commodities over long distances. (*See also* Refrigeration)

Additional reading: Schlebecker. *Whereby we thrive;* p. 171–172.

Railroad Switchyard
MECHANICAL ENGINEERING

The process of assembling freight cars into a train of 100 or 200 units requires a sorting of the cars in special areas called switchyards or hump yards. Cars are pushed to the top of a slight rise (or hump) where they are released to coast to the proper siding or track. As early as the 1850s this was used in England for sorting coal cars. Chains were used for slowing the cars. By the end of the century a special hump was built and workers had to jump onto the moving cars and stop them by hand brakes. In the 1920s hydraulic retarders were used to slow down the cars, and **punched cards** were used to identify the sidings for each car. In the 1960s in the United States computers were introduced that controlled the path and speed of each car by activating remotely controlled retarders.

Additional reading: *Illustrated science and invention encyclopedia;* v. 11, p. 1457–1460.

Ramjet. *See* **Jet Engine.**

Rankine Cycle. *See* **Steam Engine.**

RATO. *See* **Jet Takeoffs.**

Rayon
TEXTILES

The first artificial fiber to be commonly used was rayon, developed by the French chemist Louis Chardonnet in 1884. It was actually a chemically treated nitrocellulose. Rayon was highly inflammable, and Chardonnet used a process developed in 1883 by the English chemist Sir Joseph Swan for removing nitro groups from the rayon to make it far less inflammable. It led the way to development of complete synthetic fibers like **nylon.**

Additional reading: Feldman. *Scientists & inventors;* p. 286–287.

Reaper
AGRICULTURE

One of the reasons for the success of farming in the prairie states was the invention of the reaper for harvesting wheat. The first successful model was designed by the American inventor, Cyrus H. McCormick, in 1831. A better model was patented by him in 1834. A patent for another reaper had already been granted in 1833 to Obed Hussey, also an American inventor. His machine was not as good as McCormick's at reaping grain but was better at mowing hay. McCormick gradually overshadowed his rivals and went on to win a fortune as well as international fame, including being presented with the French Legion of Honor by Napoleon III. (*See also* Combine Harvester)

Additional reading: Feldman. *Scientists & inventors;* p. 144–145.

Receptor. *See* **Opioid Receptor.**

Recombinant RNA
GENETICS

In 1983 Eleanor A. Miele (U.S.), Donald R. Mills (U.S.), and Fred Russell Kramer (U.S.) successfully constructed the first recombinant RNA molecule. They introduced a technique that allowed the replication of **RNA** in large quantities.

Additional reading: Lewin. "The birth of recombinant RNA technology."

Red Blood Cell. *See* **Erythrocyte.**

Red Blood Corpuscle. *See* **Erythrocyte.**

Red Giant Star
ASTRONOMY

Red giant stars are massive and relatively cool. They evolve from stars like our sun (a yellow dwarf), which in billions of years will begin to cool, allowing its exterior to expand to have a diameter 400 times as large as it is now. It will then have become a red giant star. In time it may become smaller and more dense and develop into a **white dwarf star.** The test to determine if a star is a dwarf or a giant was found in 1914 by the American astronomer Walter S. Adams to consist of analyzing its spectrum, which he was able to do in the case of Sirius B.

Additional reading: Buedeler. *The fascinating universe;* p. 216–219.

Refraction. *See* **Light Refraction.**

Refrigeration
MECHANICAL ENGINEERING

After centuries of reliance on ice for cooling, the Scottish chemist William Cullen (1755) produced ice by causing water to evaporate in a vacuum container, and in 1834 an American inventor, Jacob Perkins, patented the use of ether as a refrigerant. Then in 1870 a German physicist, Karl von Linde, made a refrigerant from ammonia, which he began to manufacture. All inventors used the technique of alternative liquefaction and evaporation of gases. In 1930 the American chemist Thomas Midgley invented a refrigerant called Freon which was nonpoisonous and nonflammable, unlike earlier gases. The use of refrigerating freezers and **air conditioners** soon became widespread.

Additional reading: Strandh. *History of the machine;* p. 227–229.

Refueling. *See* In-flight Refueling.

Relativity
RELATIVITY

One of the best known topics of modern physics is also one that may well be the most puzzling to laymen, namely, relativity. Credit for its discovery belongs of course to the German-born Albert Einstein (U.S.), who wrote a paper in 1905 setting forth the "Special Theory of Relativity," dealing with uniform motion in a straight line. He expanded upon the **quantum theory** and also declared that nothing could have a velocity greater than that of light *(see also* Light Velocity). His paper also included what is now one of the world's best-known formulas relating energy to mass (*see also* Energy and Mass). In 1915 he wrote a paper on the "General Theory of Relativity," which brought out a new concept of **gravitation.** Tests later upheld his theories.

Additional reading: Asimov. *Asimov's new guide to science;* p. 388–394.

Relay. *See* Circuit Breaker and Fuse; Electromagnet.

Reliability. *See* Quality Control.

Reserpine
PHARMACOLOGY

Reserpine is an alkaloid extracted from certain species of the plant Rauwolfia. During the 1930s it was used in India by physicians for treating hypertension. In 1949 an Indian physician, Rustom Jal Vakil, published an account of his research with reserpine that attracted the attention of English scientists. Three Swiss chemists, J.M. Muller, Emil Schlitter, and Hugo J. Bein, isolated reserpine in 1952 and found that it was able to reduce blood pressure. Also in 1952, Robert Wallace Wilkins (U.S.) reported its sedative and tranquilizing affects, thus introducing the first tranquilizer to the West. American chemist Robert Burns Woodward synthesized it in 1956.

Additional reading: Robinson. *The miracle finders;* p. 182–185.

Reservoir
CIVIL ENGINEERING

The building of areas for holding large supplies of water, or reservoirs, was practiced by the ancient Romans, according to the writings of Vitruvius (Italy) around 30 B.C. and Frontinus (Italy) around 60 A.D. They describe how Roman water supply systems involved three reservoirs for various purposes, such as private versus public use. Centuries later reservoirs were built again, such as those built by abbeys on the outskirts of Paris around 1190. Since then reservoirs have become an indispensible part of water supply systems throughout the world.

Additional reading: Landels. *Engineering in the ancient world;* p. 25, 48–49, 51, 75.

Resin. *See* Adhesive.

Resistor. *See* Electric Resistor.

Resonance Particle
NUCLEAR PHYSICS

A group of **subatomic particles** which have extremely short lives were first discovered around 1960 by the American physicist Luis Alvarez. They have half-lives of a few trillionths of a trillionth of a second. Over 150 of them have been discovered so far and more are known to exist.

Additional reading: Asimov. *Asimov's new guide to science;* p. 361.

Reverse Transcriptase
BIOCHEMISTRY

In 1970 American oncologist Howard Martin Temin proved the existence of reverse transcriptase, an **enzyme** that directs the transcription of **RNA** into **DNA.** During the 1970s, David Baltimore, an American biochemist, discovered the same enzyme and predicted that it would affect the working of DNA. Reverse transcriptase is now a basic tool for studying the molecular biology of mammalian cells. Temin and Baltimore shared the 1975 Nobel Prize for medicine and physiology with Renato Dulbecco (U.S.).

Additional reading: Magner. *A history of the life sciences;* p. 466–468.

Revolver
ORDNANCE

There were efforts in the early nineteenth century to develop a hand gun that could quickly fire several bullets without reloading, but not until 1835 did Samuel Colt (U.S.) invent a successful model, which was first used in combat in the Mexican War around 1847. His and most succeeding models had a revolving set of barrels that would hold six bullets. A similar design was used for the Gatling guns. (*See also* Machine Gun; Silencer [Gun])

Additional reading: *Illustrated science and invention encyclopedia;* v. 5. p. 610; v. 15, p. 1981–1983.

Reynolds' Number. *See* Fluid Mechanics.

Rh Factor
IMMUNOLOGY

The Rh factor was discovered in 1940 by Austrian-born physician Karl Landsteiner (U.S.) and American physician Alexander Wiener. The name "Rh factor" comes from the fact that it was first discovered in the Rhesus monkey. Landsteiner's co-worker Philip Levine (U.S.) was the first to see the connection between Rh factor and pathology in newborn babies. He observed that problems arose when an Rh-positive fetus born to an Rh-negative mother stimulated production of anti-Rhesus **antibodies** in the mother. These antibodies created severe and frequently fatal consequences in any subsequent pregnancies experienced by the mother.

Additional reading: "20 discoveries that changed our lives"; p. 65–67.

Rhenium
CHEMISTRY

Rhenium, a very hard, platinum-white metal, atomic number 75, was discovered in 1925 by the German chemists Walter Noddack and Ida Tacke while examining **platinum** ores and columbite. It does not occur free in nature but as a minor constituent in molybdenite ores; it is commercially obtained from flue gases from the extraction of **molybdenum.** It is widely used as a catalyst in industrial reactions such as hydrogenation. Rhenium also is used as an alloy to improve ductility and tensile strength. Chemical symbol is Re.

Additional reading: *Van Nostrand's scientific encyclopedia;* p. 2446–2447.

Rhesus Factor. *See* Rh Factor.

Rhodium
CHEMISTRY

Rhodium, atomic number 45, is a member of the platinum group of metals. It was discovered in 1803 by the English chemist William Wollaston while examining some platinum ores from South America. It is almost insoluble in all acids, including aqua regia, but will dissolve in molten lead. Rhodium is found in platinum ores, especially in Siberia, South Africa, and Canada. It is used as an additive in platinum alloys, in electroplating, and in electrical conduits. Chemical symbol is Rh.

Additional reading: *Van Nostrand's scientific encyclopedia;* p. 2451.

Ribonucleic Acid. *See* RNA.

Ribosome
CYTOLOGY

In 1956 the Romanian-born biochemist George Emil Palade (U.S.) demonstrated that the site of **enzyme** manufacture in **cytoplasm** is made up of tiny particles rich in **RNA** named ribosomes. Palade was able to see these smallest of organelles using the **electron microscope,** and through his research, advanced the merger of cellular physiology with molecular biology. He received the 1974 Nobel Prize for medicine and physiology for his electron-microscopic work.

Additional reading: Judson. *The eighth day of creation;* p. 225–284, 348–446.

Richter Scale. *See* Earthquake.

Rifle
ORDNANCE

The first shoulder-controlled weapons had smooth barrels, but by 1563 rifle barrels having spiral grooves were known in Scotland. The standard United States Army rifle during the Civil War, World War I, and World War II was made by the Springfield Army Ordnance factory. Around 1860 the American inventor B. Tyler Henry invented a repeating rifle, called a carbine. The American inventor Oliver Winchester soon improved the Henry rifle. In 1937 the Army officially adopted the Garand semi-automatic rifle, using an eight-rounds clip.

Additional reading: *Illustrated science and invention encyclopedia;* v. 15, p. 1984–1986.

River Erosion
GEOLOGY

An early discussion of the nature and causes of erosion by rivers appeared in 1721, written by Henri Gautier (France). His estimates of the amount of erosion caused to rivers were based on a long career of work with canals and rivers in southern France. In 1806 Nicolas Desmarest (France) stated that lava from volcanoes flowed downhill in valleys that had been cut by rivers. Studies of the Colorado River led Clarence Dutton (U.S.) in 1862 to write about the importance of rivers as agents of erosion.

Additional reading: Faul. *It began with a stone;* p. 73, 89, 204–205.

RNA
BIOCHEMISTRY

In 1911 the Russian-born biochemist Phoebus Aaron Theodore Levene (U.S.) showed that there exist two varieties of sugar in **nucleic acid:** yeast nucleic acid containing ribose (**RNA**) and thymus nucleic acid containing deoxyribose (**DNA**). Sir Frederick Charles Bawden, an English plant pathologist, discovered in 1937 that the tobacco mosaic virus contains RNA. This was the first indication that nucleic acids are found in subcellular life as well as in all cells. In 1955 Severo Ochoa (U.S.) successfully synthesized RNA and was awarded a share of the 1959 Nobel Prize for medicine and physiology. (*See also* Messenger-RNA; Recombinant RNA; Ribosome; Transfer-RNA)

Additional reading: Gribbin. *In search of the double helix;* p. 258–294.

Road. *See* Blacktop Road; Concrete and Cement; Macadam Road.

Robot
CONTROL SYSTEMS

The concept of artificial human beings or androids has long fascinated writers and inventors. Homer's *Iliad* had a character who made servants of pure gold that were able to talk and work. During the Renaissance many inventors devised mechanical automatons, including a few attempts at androids, such as those of Hans Bullmann (Germany), who in 1550 built spring-wound figures of people who could walk and play instruments. Interest continued, and by the 1930s mechanical robots were seen in movies, fairs, etc. In 1939 Isaac Asimov, the American writer, coined the term "robotics." Creation of the field of cybernetics in 1948 by the American mathematician Norbert Wiener aided scientific work on robots. The first industrial robot was built by George Devol (U.S.) in 1954.

Additional reading: Strandh. *History of the machine;* p. 170, 175–178, 201–203.

Rocket
AEROSPACE ENGINEERING

As far back as the thirteenth century the Chinese armies used small rockets to frighten the enemy. In the nineteenth century rockets were used by European armies. In 1903 a Russian engineer, Konstantin Tsiolkovsky, wrote about rockets, but in 1926 the American engineer Robert Goddard was the first to fire a rocket using liquid fuel; later he used a guidance system and proposed multistage rockets. In 1935 he was shooting rockets faster than the speed of sound. In World War II the German rocket effort, inspired by the Romanian mathematician Hermann Oberts (who wrote on rockets in 1923), was led by Wernher von Braun. He developed the deadly V-2 rocket, used against Great Britain in 1944. Von Braun in the 1950s directed the U. S. rocket program to power vehicular **manned spaceflights.**

Additional reading: Baker. *Conquest;* p. 6–19.

Rope
MATERIALS

Rope was an early invention of man; it was used by the Egyptians in constructing the pyramids around 3000 B.C. Some Egyptian rope dating back to around 50 B.C. was found to be made of papyrus reeds. Other materials which have been used for rope include flax, jute, manila, and sisal. Around 1950 nylon and dacron began to be used. In 1754 a type of rope-making machine was invented by the English inventor Richard Marsh, followed by another English inventor, Edmund Cartwright, in 1792. (*See also* Wire Rope)

Additional reading: *Illustrated science and invention encyclopedia;* v. 15, p. 2009–2011.

Rotary Mill
AGRICULTURE

During the fifth century B.C., the first rotary mill appeared in the Mediterranean region. It was used to grind grain and was driven by either slaves or large donkeys.

Additional reading: Curwen. *Plough and pasture;* p. 128–131.

Rubber Dam
DENTISTRY

The rubber dam was invented in 1864 by Sanford C. Barnum, a New York dentist, for the purpose of keeping a tooth dry and free from saliva. Barnum used a thin sheet of oilskin and a small rubber ring to keep it around the neck of the tooth. He demonstrated the device to the New York Dental Society, and it was in wide use by 1867.

Additional reading: Ring. *Dentistry: an illustrated history;* p. 259.

Rubber Tire
MATERIALS

Although the discovery of **vulcanized rubber** took place in 1839, it was not until 1887 that the Scottish born inventor John B. Dunlop made his first pneumatic tire for a bicycle. By 1890 the pneumatic tire business was in full swing in Ireland. In time the tires were made for automobiles also.

Additional reading: Feldman. *Scientists & inventors;* p. 196–197.

Rubella Vaccine
IMMUNOLOGY

In 1961 Americans Paul D. Parkman, Malcolm S. Artenstein, and Edward L. Buescher isolated the rubella virus. Another research team led by Thomas H. Weller (U.S.) simultaneously isolated the same virus. In 1965 Parkman and Harry M. Meyer, Jr. (U.S.) successfully developed a vaccine using the live virus.

Additional reading: Robinson. *The miracle finders;* p. 54–59.

Rubidium
CHEMISTRY

A soft, silver-white metal, rubidium, atomic number 37, is a very reactive substance. Since it reacts vigorously with air and water, it is usually stored in hydrogen. It was discovered in 1861 by the German chemists Robert Bunsen and Gustav Kirchhoff during spectroscopic studies of a sample of lepidolite. One of its main uses is in **atomic clocks**, requiring little power and taking up little space. It is also used in magnetometers. Chemical symbol is Rb.

Additional reading: *Van Nostrand's scientific encyclopedia;* p. 2485–2486.

Rudder (Boat)
NAVAL ARCHITECTURE

The use of an oar to steer a boat probably began soon after the appearance of **oars,** around 3000 B.C. in Egypt. But around 200 A.D. the Chinese positioned an oar in an opening in the overhanging deck of their boats. By 1000 Arab seamen started to hang solid pieces on the stern of their ships, and by 1100 ships in western Europe began to use rudders.

Additional reading: *Illustrated science and invention encyclopedia;* v. 20, p. 5.

Ruthenium
CHEMISTRY

Ruthenium, atomic number 44, is a member of the **platinum** group of metals. It has a high melting point and is very hard. It was discovered in 1844 by the Estonian chemist Karl Klaus during a study of by-products of the refining of platinum. It is found only in association with other platinum metals. It is used as a catalyst for such processes as oxidation, hydrogenation, and isomerization. Chemical symbol is Ru.

Additional reading: *Van Nostrand's scientific encyclopedia;* p. 2486–2487.

Rutherfordium
CHEMISTRY

Rutherfordium, atomic number 104, was probably first identified in 1969 by A. Ghiorso, M. Nurmia, J. Harris, K. Eskola, and P. Eskola (all U.S.), by bombarding **californium** with carbon ions. A claim of previous discovery was made by a Soviet team led by G. N. Flerov, stating they obtained the element in 1964 by bombarding **plutonium** with neon ions. They proposed the name kurchatovium in honor of the Soviet scientist Igor Kurchatov, whereas the American name was to honor Lord Rutherford. No known use has been reported to date for the element. Chemical symbol is Rf.

Additional reading: *McGraw-Hill encyclopedia of science and technology;* v. 5, p. 1–2.

S

Safety Fuse
ENGINEERING

In order to make blasting operations safer, a slow burning fuse was developed in 1831 by the English inventor William Bickford. It is used in coal mines and also saw wartime use in land mines.

Additional reading: Giscard d'Estaing. *World almanac book of inventions;* p. 44–45.

Sail
NAVAL ARCHITECTURE

Ancient ships dating back many thousands of years were nearly all under sail, except for fighting ships, which were powered by **oars.** Until around 400 A.D. almost every sailing ship had a single square sail. In early times the ropes used to hoist the sails were made from braided ox-hide thongs. Gradually, auxiliary sails began to appear, such as a small square or an arm extending beyond the bow and a triangular one above the mainsail. Early sails were possibly made of interwoven branches, but cloth sails were known to be used by the Sumerians as early as 3500 B.C.

Additional reading: Landels. *Engineering in the ancient world;* p. 13, 154–156.

Salvarsan. *See* Arsphenamine.

Samarium
CHEMISTRY

A silver-gray color, samarium, atomic number 62, is a soft, malleable metal which must be worked and fabricated in an inert gas atmosphere. It was discovered by the French chemist Lecoq de Boisbaudran in 1879. Found mostly in monazite and bastnasite, the mineral has a light brown to hyacinth red color. It has been used in **nuclear reactors** since it will absorb **neutrons** for short periods, and also in cermets. Chemical symbol is Sm.

Additional reading: *Van Nostrand's scientific encyclopedia;* p. 2494–2495.

Satellite
AEROSPACE ENGINEERING

In 1952 the United States decided to launch manned and unmanned flights into space. In 1957 the Russians launched a satellite about four months ahead of the first United States launch in early 1958. In the next few years both countries launched satellites for studying weather conditions, measuring the solar wind, or taking photographs. For instance, in 1960 the U.S. launched the Tiros weather satellite series and in 1962 the Telstar communication satellites; both series provided much data. Satellites must generally follow an orbital path, which has many advantages, such as performing photographic missions on a continuing basis. For exploring specific targets the **space probe** serves the purpose better. (*See also* Communications Satellite)

Additional reading: Baker. *Conquest;* p. 86–102.

Saturn
ASTRONOMY

Although Saturn was known since ancient times, this bright planet was not known to have rings until the Dutch physicist and astronomer Christiaan Huygens discovered them in 1656. The Italian astronomer and physicist Galileo Galilei had noted irregularities while viewing Saturn in 1610, but his telescope was too crude to provide the detail needed. Saturn's first satellite to be discovered was found by Huygens, also in 1656. Modern **space probes** disclosed much data about Saturn, beginning in 1979, including the existence of satellites.

Additional reading: Asimov. *Asimov's new guide to science;* p. 135–140.

Scales. *See* Balance (Chemical).

Scalpel
MEDICINE

The ancient Romans used a small knife they called "scalpellus" for such surgical procedures as incisions and wound dilations. These early scalpels had handles that were either silvered or gilt, and blades made of steel.

Additional reading: Bennion. *Antique medical instruments;* p. 61–62.

Scandium
CHEMISTRY

Scandium, a relatively soft metal, atomic number 21, has a silvery luster. It was discovered in 1879 by Lars Nilson, an agricultural chemist in Sweden, while examining euxenite. It occurs in low concentration in such minerals as wolframite, wiikite, and davidite. Because of its scarcity and high cost, few applications have been found for it, although in certain light sources the use of scandium enhances luminosity. Chemical symbol is Sc.

Additional reading: *Van Nostrand's scientific encyclopedia;* p. 2524–2525.

Scanning Disk
OPTICS

The transforming of visual images so they can be transmitted was studied by many inventors over the years. A device using a revolving metal disk was invented by the Polish engineer Paul Nipkow in 1884. The disk had a number of square apertures and was placed between the subject and the light source. As the disk rotated, each part of the subject was scanned by light showing through the holes. The amount of light depended upon the shading of the subject. A **photoelectric cell** converted the light into electrical impulses. In reverse order, reproducing the pictures used electric impulses to control the intensity of the light. Nipkow patented a **television** system, but the technology to make it practical was lacking. (*See also* Television)

Additional reading: Feldman. *Scientists & inventors;* p. 238–239.

Scanning Electon Microscope. *See* Electron Microscope.

Schick Test
IMMUNOLOGY

Hungarian physician and bacteriologist Bela Schick developed the Schick test in 1909 to determine children's susceptibility to diphtheria. The test identified the presence or absence of immunity to the Coryne bacterium diphtheriae. (*See also* Diphtheria Antitoxin)

Additional reading: McGrew. *Encyclopedia of medical history;* p. 93–98.

Schmidt Camera
ASTRONOMY

In order to improve the sharpness of photographic images taken in connection with **telescopes**, the Estonian-born optician Bernhard Schmidt devised in 1930 a correcting lens that would bring rays entering near the outer edge of a lens into focus with those in the central area. His cameras have also been successfully used in spectroscopic analysis.

Additional reading: *Illustrated science and invention encyclopedia;* v. 15, p. 2046–2047.

Scientific and Technical Societies
SCIENCE AND TECHNOLOGY

Scientists and engineers derive knowledge and broaden their acquaintances with their colleagues through membership in societies organized to promote science and engineering. These groups encourage their work and give them a vehicle for publication through journals and conferences. Possibly one of the oldest and certainly one of the most prestigious is the Royal Society of London, which was chartered by Charles II (England) in 1662. Some of its early members were Robert Boyle, Robert Hooke, and Sir Isaac Newton. Many other luminaries have been members, and its "Philosophical Transactions" have been published continuously since the founding date. One of the earliest societies for engineers was the Society of Civil Engineers, also in England, founded in 1771.

Additional reading: Faul. *It began with a stone;* p. 40–42.

Scientific Exploration
SCIENCE AND TECHNOLOGY

Scientific explorations have been made since the earliest times; unfortunately few of them left written records of their findings. For example, voyages of the Portuguese Vasco da Gama in 1497 and Ferdinand Magellan in 1519 were not known for their detailed reports of their observations. However, one example of an internationally known scientist who did report fully was the German naturalist Alexander von Humboldt. His exploration of Latin America, for example, covered a multitude of disciplines, and in 1845 he published the first of an eventual five-volume work giving detailed records. He is said to have inspired the 1831 voyage of the Beagle taken by the English naturalist Charles Darwin, who built his case for evolution on data gathered in his travels.

Additional reading: Downs. *Landmarks of science;* chap. 49, 58.

Scientific Method
SCIENCE AND TECHNOLOGY

The scientific method was first used by the Greek physician Hippocrates around 400 B.C. in the treatment of disease. The Greek philosopher Aristotle also used this method around 340 B.C. in the study of biology. Great emphasis on the importance of the scientific method was given centuries later by the English scientist Sir Francis Bacon, who published a major work on the subject in 1605. He stressed the need for systematic experimentation and cooperative research.

Additional reading: Downs. *Landmarks in science;* chap. 1, 2, 23.

Scientist
SCIENCE AND TECHNOLOGY

Until the nineteenth century people engaged in research in what we know as the sciences often preferred to be called "natural philosophers" or simply "philosophers." It was not until 1833 at a meeting of the British Association for the Advancement of Science that William Whewell (English) proposed the term "scientist" for those engaged in scientific studies and experiments. Some people, such as Michael Faraday (England), disliked the narrow implication of the term; he saw himself as a natural philosopher, having a broad role to play in the world. The growth of professional societies and formal courses gradually brought around those who objected to the term.

Additional reading: Bynum. *Dictionary of the history of science;* p. 381.

Screw and Screwdriver
MECHANICAL ENGINEERING

In the sixteenth century gunsmiths found that nails in armor and guns held better if given a little twist, so they began to put slots in the nail heads, also devising small tools with a blade to turn the nails. The French inventor Andre Felibien devised in 1676 a means to unscrew the nails. Because such handmade nails were expensive, a method of making wood screws by machine was created. Around 1780 some workmen in London developed a longer-bladed screwdriver. In about 1840 L. (?) Nettlefold (England) improved wood screws by giving them points to facilitate cutting into wood.

Additional reading: De Bono. *Eureka!;* p. 226–227.

Screw Propeller. *See* **Propeller.**

Screw Threads
MECHANICAL ENGINEERING

The mechanizing of cutting screw threads was made possible by the invention of a screw-cutting lathe by the English inventor Henry Maudslay in 1810. Until 1860 each workshop had its own standards for the type and spacing of threads for its screws. As a result, it was difficult to find the proper size nut for a given size of screw. The situation was improved when the English inventor Sir Joseph Whitworth developed in 1841 a standard for screw threads, including their size and pitch. It was not immediately adopted by all manufacturers but has since been officially adopted; not all countries have the same standard.

Additional reading: Feldman. *Scientists & inventors;* p. 134–135.

Scythe
AGRICULTURE

The scythe appeared in Europe in the fifth century B.C. Its development marked a period when the weather turned colder and livestock began needing winter shelter. The scythe was used to cut hay for their feed. Two designs were predominant: 1) a short-handled scythe used in Denmark and derived from an iron **sickle;** and 2) a long-handled scythe used in Germany and Britain derived from a "balanced" scythe developed by the Romans during the Iron Age.

Additional reading: Curwen. *Plough and pasture;* p. 97, 113–121.

Sea Scale. *See* **Douglas Sea Scale; Petersen Scale.**

Seaplane
AEROSPACE ENGINEERING

The first aircraft designed to take off from and land on water was invented in 1910 by Henri Fabre (France); his plane had three flat-bottomed floats. In 1911 the American inventor Glenn Curtiss flew an amphibian plane, having both wheels and a float, allowing it to use land or water. In 1912 Curtiss demonstrated a flying boat that had a fuselage which itself floated in the water, not using extra floats. This proved to be very successful, with scheduled runs made between St. Petersburg and Tampa. In 1939 the Boeing Clipper (U.S.) began scheduled transatlantic service. In 1947 Howard Hughes (U.S.) demonstrated his huge seaplane which could have carried 700 passengers, but it was never again flown. Seaplanes are still used regularly in some isolated areas.

Additional reading: *Illustrated science and invention encyclopedia;* v. 15, p. 2054–2057.

Seawater. *See* **Chemical Oceanography.**

Seconal. *See* **Barbiturate.**

Secretin
BIOCHEMISTRY

Ernest Henry Starling and Sir William Maddock Bayliss, English physiologists, were responsible in 1902 for discovering the existence of this second known **hormone, epinephrine** being the first. It is found in the small intestine and functions as a stimulus for the pancreatic flow of digestive juices.

Additional reading: Crapo. *Hormones: the messengers of life;* p. 8–9.

Sedative. *See* Barbiturate.

Seebeck Effect
PHYSICS

One example of what is called **thermoelectricity** is the creation of an electric current in a circuit consisting of two dissimilar metals whose junctions are not at the same temperature. It was accidentally discovered in 1821 by the German physicist Thomas Seebeck. His discovery led to the invention of the **thermocouple** It is also closely related to the **Peltier effect**.

Additional reading: Feldman. *Scientists & inventors;* p. 100–101.

Seed Drill
AGRICULTURE

Jethro Tull (England) invented the seed drill in 1699 to provide a mechanism for placing seeds in the ground in small holes. The seed drill allowed for a lighter rate of seeding than did hand-broadcasting, while resulting in a higher-yield planting.

Additional reading: Hyams. *Soil and civilization;* p. 263–268.

Seismograph
GEOLOGY

The damage caused by **earthquakes** undoubtedly gave impetus to the invention of a devise for detecting and measuring them. In 1855 such an instrument was devised by the Italian physicist Luigi Palmieri. His crude invention was improved by a recording device developed in 1880 by the English engineer John Milne. He used a pen which recorded movements on smoked paper. One of Milne's contributions was his work in setting up a network of seismograph stations around the world, growing from 13 stations in 1900 to more than 500 today.

Additional reading: Asimov. *Asimov's new guide to science;* p. 162–165, 169–171.

Selenium
CHEMISTRY

Selenium, atomic number 34, was first identified by the Swedish chemist Jons Berzelius in 1817. Although found in many sulfide ores, selenium is obtained by treating by-products of copper refineries. It is widely used in photoelectric cells, in electric rectifiers, and in paint and glass products. Chemical symbol is Se.

Additional reading: *Van Nostrand's scientific encyclopedia;* p. 2542–2544.

Self-starter. *See* Automobile Self-starter.

Semiconductor
SOLID-STATE PHYSICS

Materials having electrical properties between conductors and insulators are called semiconductors. Some scientists at Bell Laboratories in the 1940s discovered a careful mixing of impurities enhanced the conductivity of semiconductors. Studies made in the 1950s of semiconductor properties brought the Nobel Prize to Sir Nevill Francis Mott (England) and Philip Warren Anderson (U.S.) Several major devices are based on semiconductors, such as **transistors** and the tunnel diode. The latter, invented in 1957 by Lee Esaki (Japan), became the basis for switching devices. Thermistors are semiconductors used for measuring temperatures (based on measuring their electrical resistance).

Additional reading: *Illustrated science and invention encyclopedia;* v. 15, p. 2075–2077.

Servomechanism
CONTROL SYSTEMS

Servomechanisms are devices which act as automatic control systems governing mechanical motion, featuring a small input of power to control a large output of power. The first example was a governor for a steam engine devised in 1775 by James Watt (England) to control the speed of the engines. In 1868 the French engineer Leon Farcot invented a device for controlling a steamship's rudder; he called his invention a servomechanism. With the advent of electronics decades later, the variety of servomechanisms invented increased greatly. An example of modern types is the automatic pilot for **aircraft.** One feature of these devices is the principle of using negative feedback, resulting in greater stability of the system.

Additional reading: *Illustrated science and invention encyclopedia;* v. 15, p. 2077–2080.

Sewer. *See* Waste and Sewage Disposal.

Sewing Machine
MECHANICAL ENGINEERING

A device with both domestic and commercial applications is the sewing machine. The first patent was given in 1790 to Thomas Saint, an English cabinetmaker. One of the earliest commercial models was invented in 1830 by the French tailor Barthelemy Thimonnier, but angry tailors and seamstresses destroyed the machine, fearful of losing employment. In 1846 the American engineer Elias Howe invented a two-thread, lock-stitch machine. He won a patent suit in 1854 against the American engineer Isaac Singer, who had added improvements in 1851, such as the foot treadle, improving its use for workshops. Later models have been electrified and given automatic controls; the Pfaff Company (Germany) brought out an electronic model in 1968.

Additional reading: *Illustrated science and invention encyclopedia;* v. 15, p. 1084–1087.

Sex Hormone
BIOCHEMISTRY

French physiologist Charles Edouard Brown-Sequard injected himself in 1889 with several doses of guinea pig testicle extract. After noticing sensations of well-being, he hailed his treatment as a rejuvenation. Although his work in this area was soon discredited, Brown-Sequard did advance the development of sex hormone research.

Additional reading: Sneader. *Drug discovery;* p. 192–195.

Sex-Linkage Principle
GENETICS

In 1910 the American geneticist Thomas Hunt Morgan obtained a mutant white-eyed male Drosophila in a colony of all red-eyed Drosophila. After proceding with further breeding experiments, he had a mix of red-eyed females and white-eyed males. As a result of these experiments, he developed the principle of sex-linkage, wherein certain hereditary factors are linked to either the male or the female **chromosome**. Morgan was awarded the 1933 Nobel Prize in medicine and physiology for his work "concerning the function of the chromosome in the transmission of heredity."

Additional Reading: Dunn. *A short history of genetics;* p. 104–109, 139–147.

Shear Modulus. *See* Modulus of Elasticity.

Ship
NAVAL ARCHITECTURE

Ships have been used by man since the earliest times. It is known that the Egyptians sailed in them as early as 4000 B.C. Whether powered by **sail** or by **oars**, ships enabled exploration and trade to take place all over the world. Designs of ships changed constantly, as skills and materials improved. For example, the relatively speedy American clipper ships, first built in 1845, set new records for sailing vessels to the Far East. Speed records for motor-powered **ocean liners** were continually being bested. With the advent of iron hulls in ships in the 1840s, safety became more of a factor in ship design. Modern navigation aids became common after World War II. (*See also* Boat; Oar)

Additional reading: *Encyclopedia Britannica;* 15th ed., v. 18, p. 663–674.

Sickle
AGRICULTURE

Evidence of the earliest sickle was found in caves in Mount Carmel in Palestine and dated to some time before 6000 B.C. It was probably used by Mesolithic hunters and gatherers to cut grasses or grains, and was made of flint mounted on carved bone handles.

Additional reading: Curwen. *Plough and pasture;* p. 103–121.

Silage
AGRICULTURE

Silage, fodder that is fermented to reduce spoilage, was introduced when the first silo was built in the United States in 1873. Grasses, grains, and legumes could be stored for a long time in silos and were enhanced by biochemical reactions that added nutritional value and palatability to the fodder. By the 1890s most dairymen were producing silage in silos, and were able to winter over more cattle as a result.

Additional reading: Schlebecker. *Whereby we thrive;* p. 183.

Silencer (Gun)
ORDNANCE

The firing of a hand gun can be silenced if the gases escaping from the muzzle can be slowed down. A silencer to accomplish this was patented in 1908 by the American inventor Percy Maxim. His device was a cylinder attached to the gun's muzzle having a series of chambers separated by baffles with holes in them The escaping gases were slowed down after the firing of the gun. Silencers are most effective with low velocity ammunition.

Additional reading: *Illustrated science and invention encyclopedia;* v. 16, p. 2105–2106.

Silicon
CHEMISTRY

Silicon, atomic number 14, is a dark-gray, hard solid. It can also be obtained in a dark-colored crystalline form. In the early 1800s several chemists tried unsuccessfully to isolate silicon, but the Swedish chemist Jons Berzelius was able to do so in 1824 by treating silicon tetrafluoride. Silicon does not appear in elemental form in nature, always existing in combined form in rocks, minerals, and seawater. It is used in steel production, as an industrial abrasive, and, in recent years, in **semiconductors**. For the latter use the silicon must be in a superpure state (only one atom of impurity is allowed for every 100,000 silicon atoms). Over 800 tons of this form is used annually by the semiconductor industry. Chemical symbol is Si.

Additional reading: *Van Nostrand's scientific encyclopedia;* p. 2570–2574.

Silo. *See* **Silage.**

Silver
CHEMISTRY

A white metal, silver, atomic number 47, is softer than **copper** and harder than **gold**. It is superior to all other metals as a conductor of electricity and heat. Slag dumps in Asia Minor show that silver had been separated from lead as early as 3000 B.C. Although silver is found all over the world, it exists in some areas as a compound such as silver chloride. Most silver is obtained as a by-product from treatment of **lead**, copper, gold, etc. Besides use in jewelry and coinage, silver is used in industry for photographic film, electrical contacts, tableware, and batteries. Chemical symbol is Ag.

Additional reading: *Van Nostrand's scientific encyclopedia;* p. 2575–2577.

Silver Mining
MINING

Silver was a medium of exchange in Assyria as far back as 700 B.C. Slaves were widely used in Greek silver mines around 600 B.C. Donkey-driven mills were being employed at that time to crush the ore during processing. By the Middle Ages, central Europe saw a large increase in the amount of silver being mined, particularly in Hungary and parts of Germany. Rich deposits were mined in Bohemia around 1516, techniques of which were fully described in the famous book *De Re Metallica,* by Georgius Agricola in 1556. By 1623 German miners had set up silver mines in Norway. In 1833 a new method for extracting silver from **lead** was patented by H. L. Pattinson (England), and in 1865 J. B. Elkington (England) patented the electrolytic refining of **copper**, from which silver was obtained.

Additional reading: Derry. *Short history of technology;* p. 12, 34, 39, 122–123, 249, 488, 493.

Sleeping Pill. *See* **Barbiturate.**

Slide Rule
MATHEMATICS

For several centuries the slide rule has been a most popular type of portable calculator, capable of handling simple calculations as well as complicated ones in models designed for professional use. The slide rule is based on the use of logarithmic scales. It was invented in 1621 by the English mathematician William Oughtred, although one of his pupils, Richard Delamain, another English mathematician, claimed to have invented it by virtue of a paper he wrote in 1620 on the circular slide rule. In recent years slide rules have all but disappeared from use as electronic calculators have taken their place. (*See also* Logarithm)

Additional reading: *Illustrated science and invention encyclopedia;* v. 16, 2123–2125.

Slit-Lamp
OPHTHALMOLOGY

The slit-lamp was developed in the 1890s and 1900s by Swedish physician Allvar Gullstrand. The slit-lamp supplied a brilliant light which was condensed into a beam that illuminated parts of the eye for purposes of examination. When used in conjunction with a binocular microscope, the slit-lamp allowed the examiner to identify inflammation or locate foreign bodies deep in the eye.

Additional reading: Sourkes. *Nobel prize winners in medicine and physiology 1901–1965;* p. 69–73.

Smallpox Vaccine
IMMUNOLOGY

In 1715 Giacomo Pylarini (Italy) described a smallpox inoculation he had given three children in 1701. In 1796 English physician Edward Jenner discovered a vaccine that used active cowpox to effectively immunize against smallpox. Jenner successfully inoculated an eight-year-old boy, James Phipps. Jenner named this inoculation "vaccination" from the Latin word "vaccinia," meaning cowpox. Jenner's method was much safer than earlier vaccines such as Pylarini's. (*See also* Immunization)

Additional reading: Feldman. *Scientists & inventors;* p. 78–79.

Societies. *See* **Scientific and Technical Societies.**

Sodium
CHEMISTRY

Sodium, atomic number 11, is a silvery-white metal that reacts instantly in air and water, thus needing storage in kerosene. The English chemist Sir Humphry Davy was the first to isolate sodium (1807), using **electrolysis.** A cheaper way of producing sodium was developed independently in 1854 by the German chemist Robert Bunsen and the French chemist Charles Sainte-Claire Deville in an experiment to prepare pure aluminum. Because sodium is so reactive, it does not occur in nature in the free state, but appears as salt in the ocean, in salt lakes, salt dunes, etc. It is used as an alloying constituent, in sodium vapor lights, and in electrical conductors. Chemical symbol is Na.

Additional reading: *Van Nostrand's scientific encyclopedia;* p. 2600–2604.

Sodium Barbital
PHARMACOLOGY

In 1903 German chemist Emil Hermann Fischer filed for a patent on diethylbarbituric acid, a powerful new hypnotic drug isolated under his direction. Also known as sodium barbital, F. Bayer and Company marketed it as Veronal. (*See also* Barbiturate)

Additional reading: Sneader. *Drug discovery;* p. 24–34.

Solar Battery
ELECTRONICS

A means for converting sunlight directly to electricity is the solar battery, sometimes called a photovoltaic cell. It contains **semiconductors** which readily lose **electrons** when exposed to sunlight. **Silicon** batteries were invented in the United States in 1954 by D. M. Chapin, C. S. Fuller, and G. L. Pearson (all U.S.) at Bell Laboratories. They have proved useful in space vehicles and **space probes**.

Additional reading: Asimov. *Asimov's new guide to science;* p. 463–464.

Solar Energy
MECHANICAL ENGINEERING

Harnessing the energy of the sun is by no means a modern invention. The ruins of buildings in Nineveh contained lenses and mirrors that are believed to have been used around 600 B.C. to focus rays of the sun. A relatively modern application came in 1883 when the American engineer John Ericcson built a "sun motor" which could heat a boiler. Some solar heaters on a small scale were built before World War II, but the most important project is located in France at Odeillo, high in the Pyrenees, where a 130-foot high mirror was built around 1950. It contains 9,500 mirrors, focused on a solar oven, where temperatures of nearly 7,000 degrees Fahrenheit have been obtained. Many smaller, mostly residential projects are currently being carried on.

Additional reading: Strandh. *A history of the machine;* p. 163–166.

Solar Heating
MECHANICAL ENGINEERING

The rising costs of conventional fuels spurred the recent development and use of solar heating. A small solar device that could run a **steam engine** was demonstrated in Paris in 1878, built by the French inventor Mouchot. Research into solar heating was not active until the 1940s, later increasing dramatically during the energy crisis in 1973. Solar collectors absorb energy and transfer it to tanks of water which circulate in the central heating system. By reversal of the equipment it can also be used to cool a building. Hot water heaters can also be run on solar energy. Usually a conventional heating system is needed as a backup for solar heating.

Additional reading: *Illustrated science and invention encyclopedia;* v. 16, p. 2134–2138.

Solar System
ASTRONOMY

Various astronomers have proposed many models for the relative position of the sun, earth, and planets. For over fourteen centuries the world accepted the theory of Claudius Ptolemy, a Greek astronomer, who stated around 150 A.D. that the earth is a fixed body at the center of the universe, around which the sun and fixed stars revolve. This belief was not challenged until a Polish astronomer, Nicolaus Copernicus, prepared a manuscript in 1530 which stated that the sun was the center around which all planets moved. He was supported by the noted Italian astronomer Galileo Galilei, whose book in 1632 was bitterly opposed by church authorities. Further support came from Johannes Kepler, a German astronomer, who wrote on **planetary motion** in 1609.

Additional reading: Downs. *Landmarks in science;* chap. 10, 15, 24.

Solar System's Origin
ASTRONOMY

In 1745 the French naturalist George Buffon suggested that the earth was created by the collision of a comet with the sun. In 1786 the French astronomer Pierre Laplace proposed that the sun originated from a rotating cloud of gas which threw off rings of material. The rings would then condense and form planets, a theory discredited in 1859 by the Scottish physicist James Clerk Maxwell. In 1905 the American scientists Forest Moulton and Thomas Chamberlin proposed that a passing star was responsible for drawing the material which later condensed to form planets. Other major theorists have included Lyman Spitzer, Jr. (U.S.) in 1939, Carl von Weizsacker (Germany) in 1944, and Sir Fred Hoyle (England) in 1946. The controversy continues. (*See also* Earth's Formation; Universe's Origin)

Additional reading: Asimov. *Asimov's new guide to science;* p. 95–101.

Solar Wind
ASTROPHYSICS

Studies of the solar atmosphere in the 1920s led the English physicist Edward Milne to realize that the sun ejects particles, some at speeds of up to 1,000 kilometers per second. In the 1930s the Italian-born physicist Bruno Rossi (U.S.) used rockets to study cosmic rays, and he became aware of the streams of particles flowing from the sun in all directions. In 1959 the American physicist Eugene Parker predicted that charged particles would be emitted from the sun, following the lines of force of its magnetic fields. This was verified in 1962 when the Mariner 2 Venus probe was launched. By now known as the "solar wind," this phenomenon helped explain why comets' tails point away from the sun (the solar wind strikes the haze surrounding a comet and sweeps it outward in a tail).

Additional reading: Asimov. *Asimov's biographical encyclopedia;* p. 750–751, 804, 886.

Soldering
METALLURGY

As far back as 2500 B.C. the Chaldeans used a form of soldering to join gold sheets. Solder is a metal alloy with a melting point lower than that of the metals to be joined. In 1838 the burning of lead with hydrogen was proposed by the French inventor Desbassaynes de Richemond. Electric soldering irons were developed years afterward, allowing for more precise operations.

Additional reading: *Illustrated science and invention encyclopedia;* v. 16, p. 2138–2139.

Sonar
ACOUSTICS

The use of sound waves for underwater submarine detection was investigated in 1912 at the suggestion of the English physicists Lord Rayleigh and Owen Richardson. That year in France a Russian engineer, Constantin Chilovsky, prepared a model detector, but it failed. Around 1917 the French physicist Paul Langevin succeeded; both had used piezoelectricity to create ultrasonic waves. By World War II the system was perfected, becoming known as sonar, which is an acronym standing for "Sound Navigation And Ranging." Besides submarine detection, sonar can also be used for mapping ocean bottoms, detecting schools of fish, etc. The equipment includes a **transducer** for creating the sound waves and a cathode ray screen for displaying results. (*See also* Ultrasonics)

Additional reading: *Illustrated science and invention encyclopedia;* v. 2, p. 157–158.

Sound Recording. *See* Magnetic Recording; Phonograph; Stereophonic Sound.

Space Probe
AEROSPACE ENGINEERING

Space probes, thought by many to be more effective ways of gathering data from space than **manned spaceflights**, were some of the earliest types of space objects launched. Unlike **satellites,** having only an orbital path, space probes can be directed to one or more targets in space. The first one, launched by the Soviet Union in 1959, was a lunar probe, sent to measure radiation and ionization near the moon; it was followed in nine months by one which hit the moon. In the 1960s the U.S. launched several successful probes deep into space, designed to study data from various planets on trips lasting, in some cases, several years and eventually going beyond the solar system. Photographs and numerical data have been obtained over the years from these flights.

Additional reading: Baker. *Conquest;* p. 103–119.

Space Ship. *See* Manned Spaceflight; Space Station.

Space Shuttle. *See* Manned Spaceflight.

Space Station
AEROSPACE ENGINEERING

In recent years the United States and the Soviet Union have both made manned space vehicles capable of sustaining humans in orbit for longer and longer time periods. The United States launched its most ambitious vehicle, Skylab, in 1973. Three different crews manned it for a total of 170 days, having taken over 150,000 photographs and over 45 miles of magnetic tape. In 1982 two Soviets manned Soyuz 7 for a record-breaking 211 days. Plans for a permanent space station have been considered in the 1980s by both countries but implementation is uncertain even though designs for those projects go ahead. Even more esoteric are plans for a permanent space settlement where as many as 10,000 people might take up residence.

Additional reading: Baker. *Conquest;* p. 120–141, 169–171.

Sparkplug. *See* **Automobile Ignition System.**

Spectacles. *See* **Eyeglasses; Lens.**

Spectroscope
OPTICS

A study of the visible spectrum in the light emitted from heated chemicals led the German physicist Gustav Kirchhoff and the German chemist Robert Bunsen to invent the spectroscope in 1859. They caused the light to pass through a narrow slit, resulting in images of the slit representing the various wavelengths of the light being studied. This was the beginning of chemical analysis by spectroscopy, and using the method enabled them to discover the two elements **cesium** and **rubidium.**

Additional reading: Feldman. *Scientists & inventors;* p. 168–169.

Spectroscopic Binary Star. *See* **Binary Star.**

Spectroscopy
OPTICS

Spectroscopy is the study of radiation by determining its component frequencies and the relative intensity of each. Although more frequently concerned with **light** and **X-rays**, it can also be applied to elementary particles. Sir Isaac Newton (England) made some of the earliest studies in 1704 with his prisms dispersing light into its constituent colors. In 1814 **Fraunhofer lines** were discovered. The development in 1859 of the recording **spectroscope** made spectroscopy a suitable method of chemical analysis. Around 1912 H. E. J. Mosely (England) developed X-ray spectroscopes using equipment in which a crystal replaced a diffraction grating or a prism.

Additional reading: Bynum. *Dictionary of the history of science;* p. 397.

Speech Recognition
COMPUTER SCIENCE

In 1926 a scientist at Bell Laboratories (U.S.) developed the voice coder (or vocoder), which analyzes the pitch and energy content of speech. In time the output of the device was put in digital form. In 1950 H. K. Davis (U.S.) invented a system that could distinguish a series of ten spoken numbers. By 1971 some systems developed in the United States had a vocabulary of around 1,000 words. By 1983 a Japanese device could distinguish 1,000 words with an accuracy of 99%. Continued progress has been made since then.

Additional reading: Scott. *Introduction to interactive computer graphics;* p. 23–25.

Sperm
HISTOLOGY

Sperm were first observed in 1677 by the Dutch biologist and microscopist Anton van Leeuwenhoek with the aid of his single lens **microscope.** In 1779 Lazzaro Spallanzani (Italy) showed that sperm cells needed to make contact with egg cells before **fertilization** could occur. Rudolf Albert von Kolliker, a Swiss anatomist and physiologist, demonstrated in 1841 that sperm are sex cells that result from a transformation of cells in the testis.

Additional reading: Magner. *A history of the life sciences;* p. 161, 194–196.

Sphygmomanometer
MEDICINE

In 1863 French physiologist Etienne Jules Marey invented the first sphygmograph for the purpose of recording blood pressure. An external sphygmomanometer that allowed the measurement of blood pressure in clinical settings was developed in 1896 by Scipione Riva-Rocci (Italy). (*See also* Kymograph)

Additional reading: Woglom. *Discoverers for medicine;* p. 11–29.

Sphygmometer
MEDICINE

The sphygmometer was invented in 1835 by French physician Julius Herisson. It transmitted impulses from the pulse beat to a mercury column and made each beat visible to the observer. Herisson's sphygmometer was the first tool to visually show and numerically measure the pulse beat without the need to puncture an artery. (*See also* Pulse Watch)

Additional reading: Reiser. *Medicine and the reign of technology;* p. 98–100.

Spinning Jenny
TEXTILES

The production of yarn by spinning wheels was a slow process. It was speeded up by the invention around 1764 of the spinning jenny by the English weaver James Hargreaves, whose invention could do the work of 30 spinning wheels. The invention in 1779 of the spinning mule by Samuel Crompton, an English inventor, produced yarn of a higher quality than that from Hargreaves' machine. One operator could spin up to 1,000 threads. (*See also* Water Frame)

Additional reading: Feldman. *Scientists & inventors;* p. 56–57.

Spinning Mule. *See* **Spinning Jenny.**

Spiral Nebula. *See* **Galaxy.**

Spirometer
MEDICINE

John Hutchinson, an English physician, built the spirometer in the 1840s so that he could measure the lungs' vital capacity. The spirometer consisted of a tube through which a patient breathed in order to elevate a receiver to varying levels that reflected lung capacity. Hutchinson demonstrated that vital capacity changed according to a person's state of health or disease.

Additional reading: Reiser. *Medicine and the reign of technology;* p. 91–95.

Spirophore. *See* **Iron Lung.**

Spray Can. *See* **Aerosol Spray.**

Stabilizer. *See* **Gyrostabilizer.**

Stained Glass. *See* **Glass.**

Stainless Steel
MATERIALS

Corrosion-resistant materials have many uses. An ideal type is an alloy that does not corrode. A major alloy was found accidentally in 1913 by the British metallurgist Harry Brearley. After examining several alloys and discarding them, he noticed months later one in particular, a nickel-chromium alloy, was still bright and shiny in a scrap heap. It became known as stainless steel, an excellent corrision-resistant metal.

Additional reading: Asimov. *Asimov's new guide to science;* p. 312.

Stamen. *See* **Plant Anatomy.**

Star Catalog
ASTRONOMY

The first known map of the position of the stars was made by the Greek astronomer Hipparchus around 130 B.C. His work was later extended by Claudius Ptolemy, another Greek astronomer, around 150 A.D., a copy of whose work somehow escaped destruction and came to light in 765 A.D. One of the first astronomers to keep accurate records of the positions of stars was the Danish scientist Tycho Brahe. Using only the naked eye for his observations, he published a large catalog in 1592, which was still incomplete at the time of his death in 1601. He also studied the orbits of the planets and many other phenomena.

Additional reading: Asimov. *Asimov's biographical encyclopedia;* p. 33–35, 42–43, 91–95.

Star Wars. *See* **Laser.**

Static Electricity. *See* **Electricity.**

Steam Engine
MECHANICAL ENGINEERING

The first practical machine run by steam was the **steam pump.** The work of one man, the Scottish inventor James Watt, caused great changes in the design and use of steam engines. In 1769 he received a patent for an engine which reduced operating costs. In 1781 he patented one that would provide rotary rather than reciprocating action, thus widening applications, such as running spinning and weaving equipment. He also invented a double-action engine and a governor (to keep the speed constant). Use of steam at high pressure was developed by Richard Trevithick, an English inventor, in 1801. In 1859 the thermodynamics theory of steam engines was described by William Rankine (Scotland). (*See also* Steam Locomotive; Steam Turbine)

Additional reading: Clark. *Works of man; p. 72–80; 112–114.*

Steam Hammer
MECHANICAL ENGINEERING

The manufacturing of metal objects by a pounding or forging process was simplified by the invention of the steam hammer in 1839 by the Scottish engineer James Nasmyth. It was a huge piece of equipment, several stories high, making it possible to produce mammoth forgings. It was the direct predecessor of the **pile driver.**

Additional reading: Feldman. *Scientists & inventors;* p. 138–139.

Steam Heating
MECHANICAL ENGINEERING

A steam heating system consists of a boiler which converts water to steam, which then rises through pipes to radiators. The Scottish inventor James Watt studied steam heating in 1784 and a system was installed in his factory near Birmingham. Another system was patented in England in 1791. It had the advantage of being a clean system but was not able to make quick accommodation to changing temperatures.

Additional reading: *Illustrated science and invention encyclopedia;* v. 9, p. 1187–1191.

Steam Locomotive
MECHANICAL ENGINEERING

The development of railroads was dependent upon the invention of adequate locomotives. The first one was built by the English engineer Richard Trevithick in 1804, but it proved to be too heavy for the crude rails of the time. Others were built for use only around coal mines. The first practical locomotive was devised in 1825 by the English engineer George Stephenson. His main contribution was the invention of the steam-blast engine in which the waste steam was routed through a narrow pipe that improved the fire in the firebox. This enabled the locomotive to pull trains at speeds of up to 14 miles per hour. In later years other improvements were made in the valves and in the use of superheated steam. (*See also* Diesel Engine)

Additional reading: *Illustrated science and invention encyclopedia;* v. 11, p. 1394–1399.

Steam Pump
MECHANICAL ENGINEERING

The creation of steam-driven devices could not have come until the skills existed for making cylinders of brass or cast-iron and for building sturdy boilers (in which steam is created) from iron. These products depended upon **iron smelting.** The first successful use of steam to pump water was the device patented in 1698 by Thomas Savery, an English military engineer. It was used to drain water from mines. Later improvements were made by two other Englishmen, John Smeaton (around 1750) and Thomas Newcomen in 1712. (*See also* Steam Engine)

Additional reading: Clark. *Works of man;* p. 60–82.

Steam Turbine
MECHANICAL ENGINEERING

A means of providing power that was to result in a plethora of applications is the steam turbine. It was invented in 1884 by the English engineer Sir Charles Parsons. Knowing the limitations of the **steam engine,** he devised a turbine, which had a rotor in which vanes were attached. Steam striking the vanes would cause the shaft to revolve. This design increased the speed of operation from 1,500 to 18,000 revolutions per minute. In time steam turbines were to operate in **electric power plants** for a city as well as **ocean liners.**

Additional reading: Feldman. *Scientists & inventors;* p. 220–221.

Steamboat
NAVAL ARCHITECTURE

The history of the development of a steam-driven ship goes back to the period when steam engines were being perfected. Around 1780 the American inventor John Fitch built a series of steamboats. His second model made regular trips on the Delaware River but was not a financial success. The credit for building the first financially successful steamship goes to another American inventor, Robert Fulton, whose *Clermont* was launched in 1807. He also built the first steam-powered warship (the *Demologos*) which was launched in 1814.

Additional reading: Feldman. *Scientists & inventors;* p. 90–91.

Steel
METALLURGY

Steel is undoubtedly the world's most important alloy. It has a long history, having been made in small quantities for over 2,000 years. In 1740 Benjamin Huntsman (England) found a way to produce steel of a rather homogeneous nature, involving the melting of wrought iron bars. A major event in the history of steel was the discovery in 1856 of a converter by Sir Henry Bessemer (England). It produced steel in large quantities at a cheaper cost than previously. Steel production grew quickly after that. Other methods were subsequently adopted over the years. (*See also* Basic Oxygen Process; Bessemer Converter; Electric Arc Process; Open Hearth Process)

Additional reading: *Illustrated science and invention encyclopedia;* v. 17, p. 2238–2241.

Stellarator
PLASMA PHYSICS

A magnetic container that will hold hot gases known as plasma has been difficult to design. A figure-eight shape, known as the stellarator, has been tried. It was designed in 1951 by the American physicist Lyman Spitzer, Jr. (*See also* Plasma)

Additional reading: Asimov. *Asimov's new guide to science;* p. 503.

Stereophonic Sound
ACOUSTICS

The first serious study of stereophonic recording took place at Western Electric and the Bell Telephone Laboratories in the 1920s. In 1931 the latter organization demonstrated stereophonic sound. Also in 1931 the English physicist A. D. Blumlein developed a stereophonic system. The first stereo records were produced in 1933 by the British firm EMI (Electronics and Musical Industries). Within a decade or so radio broadcasting, phonographs, and audio tape players provided stereophonic sound at a cost within reach of the general public.

Additional reading: *Illustrated science and invention encyclopedia;* v. 17, p. 2248–2254.

Stethoscope
MEDICINE

Theophile Rene Hyacinthe Laennec, a French physician, invented the first stethoscope in 1816. Laennec's instrument was basically a paper tube that enabled the physician to hear and interpret the patient's heart beat. Laennec designed a wooden stethoscope soon after he constructed the original paper tool.

Additional reading: Bettmann. *A pictorial history of medicine;* p. 238–239.

Stock Ticker
ELECTRICAL ENGINEERING

In 1869 a breakdown of a gold-price telegraphic indicator in a Wall Street office brought in Thomas Alva Edison, the American inventor, to repair it. He did so and stayed on to invent an effective printer of stock prices, the familiar stock ticker.

Additional reading: Feldman. *Scientists & inventors;* p. 206–207.

Storage Battery. *See* **Electric Battery.**

Stove
MECHANICAL ENGINEERING

The cooking stove gradually evolved from the fireplace. In the 1630s John Sibthorpe (England) patented an oven to be used with coal, but not until 1780 did Thomas Robinson of England invent the first kitchen stove. It had a cast iron oven on one side of a fire grate. A popular type was designed by Benjamin Franklin (U.S.) in 1744 and the "Franklin stove" was widely adopted in Europe several decades later. In 1802 a closed-top cooking range was invented by George Bodley (England). An American version, a freestanding unit, had legs and an iron flue. It became the model for **gas stoves** and **electric stoves.**

Additional reading: *Encyclopedia of inventions;* p. 34.

Strange Particle
NUCLEAR PHYSICS

After creating the concept of quarks in 1964, American physicist Murray Gell-Mann proposed that there would have to be a third kind of quark in addition to his original supposition of only two types. He called the third one the s-quark. The "s" stands for "strangeness" and refers to the strange nature of these particles. They existed for much longer times before breaking down than particles that were not "strange."

Additional reading: Asimov. *Asimov's new guide to science;* p. 361, 363, 364.

Strategic Defense Initiative. *See* Laser.

Stratosphere
METEOROLOGY

In 1902 the French meteorologist Leon Teisserene de Bort proposed that the atmosphere is divided into two parts, with the lower layer containing all the weather activity (such as rain, clouds, winds). He called this the troposphere, and the higher layer he named the stratosphere, beginning at 25,000 feet at the poles and 50,000 feet at the equator. The boundary between the two he called the tropopause.

Additional reading: Asimov. *Asimov's biographical encyclopedia;* p. 862–863.

Streptomycin
MICROBIOLOGY

The **antibiotic** streptomycin was isolated from a soil mold of the genus Streptomyces in 1943 by Russian-born Selman Abraham Waksman (U.S.). In 1945 physicians associated with the Mayo Clinic in Rochester, Minnesota, administered it to a patient suffering from tuberculosis. The patient survived, and streptomycin became the preferred method of treatment until the introduction of **Isoniazid.** Waksman received the 1952 Nobel Prize in medicine and physiology for his discovery.

Additional reading: Robinson. *The miracle finders;* p. 4–9.

Stress and Strain. *See* Modulus of Elasticity.

Stroboscope
ILLUMINATING ENGINEERING

A lamp which produces flashes of light at regular, controlled intervals is called a stroboscope. It was invented in 1833 by Simon Von Stampfer (Austria). At the same time the Belgian physicist Joseph Plateau invented what he called a phenokistiscope, which also involved flashing lights. Most stroboscopes now use a xenon-filled discharge lamp.

Additional reading: *Illustrated science and invention encyclopedia;* v. 17, p. 2280–2281.

Strontium
CHEMISTRY

Strontium, atomic number 37, is a silver-white metal which is soft as lead, ductile, and malleable. It was discovered in 1790 by the Irish physician Adair Crawford. The metallic element was isolated in 1808 by the English chemist Sir Humphry Davy. It occurs as a sulfate or carbonate, and the best sources are found in England. The largest use is in flares, fireworks, and tracer bullets as well as lubricants and luminescent paint. Chemical symbol is Sr.

Additional reading: *Van Nostrand's scientific encyclopedia;* p. 2695.

Subatomic Particle
NUCLEAR PHYSICS

The discovery of the **electron** in 1897 by the English physicist Joseph John Thomson opened the door to the continuing search for new subatomic particles. Some took years to discover; others have yet to be discovered. The need for high energies to produce some of them led to the invention of the **particle accelerator** in various forms. In addition some so-called particles were found to be radiation; often the distinction took much study to determine. (*See also* names of specific particles and radiations)

Additional reading: Asimov. *Asimov's new guide to science;* p. 317–372.

Submarine
NAVAL ARCHITECTURE

A treatise written in 1580 by the English naval officer William Bourne about underwater ballasting showed that a ship could be submerged and return to a floating position by taking in or expelling ballast water. In 1624 Cornelius van Drebbel (Holland) built what is said to be the first submarine. In 1776 David Bushnell (U.S.) built a hand-cranked submarine. In 1799 Robert Fulton (U.S.) built in France a propeller-driven submarine, but its tests failed to impress Napoleon to give it further backing. In 1900 the U.S. Navy purchased several submarines based on a model built by John Holland (U.S). Propulsion later came from steam, **electric motors,** gasoline motors, and in 1955 the first nuclear submarine was built under the direction of Admiral Hyman Rickover (U.S.).

Additional reading: *Illustrated science and invention encyclopedia;* v. 17, p. 2285–2289.

Submarine Cable
COMMUNICATIONS

The first submarine cable for transatlantic service was laid in 1858 under the direction of the Englishman Charles Bright. It was insulated with gutta percha and protected with galvanized iron wires; it was used for telegraph communication. The first cable was damaged that same year, and not until 1867 was another cable installed, this one under the supervision of Cyrus Field (U.S.). Beginning in 1921 coaxial cables were introduced and telephone service was later provided, starting in 1956 to Europe and in 1957 to Hawaii. The use of repeaters at suitable intervals designed to amplify the signals had begun in 1943. In 1961 a lightweight cable having repeaters every 26 miles was first used. Modern cable-laying ships can operate at a speed of eight knots.

Additional reading: *Illustrated science and invention encyclopedia;* v. 4, p. 409–412.

Succession
ECOLOGY

In 1905 F. E. Clements (U.S.) introduced the concept of succession in his book entitled *Research Methods in Ecology.* He explained succession as a phenomenon in which new species invade an area and cause the disappearance or decrease of the original occupants. In 1916 Clements published *Plant Succession* and stated a universal law that "all bare places give rise to new communities...."

Additional reading: McIntosh. *The background of ecology;* p. 79–85.

Sulfanilamide
PHARMACOLOGY

German biochemist Gerhard Domagk was looking for chemical substances that could kill bacteria when, in 1932, he discovered that a red dye named Prontosil effectively protected mice against streptococcal infections. In 1936 Jacques Trefouel (France) and Daniele Bovet (France) broke Prontosil down to its effective fragment, which was sulfanilamide. These discoveries introduced the first of the sulfa drugs, which soon came to be regarded as "wonder drugs." Domagk was awarded the 1939 Nobel Prize in medicine and physiology for his discovery. The introduction of other sulfa drugs such as sulfapyridine, sulfathiazole, and sulfadiazine followed in 1937, 1939, and 1941, respectively.

Additional reading: Asimov. *Asimov's new guide to science;* p. 658–659.

Sulfur
CHEMISTRY

One of the few elements found in the earth in elemental form, sulfur, atomic number 16, was known as far back as the time of Homer (around 800 B.C.). In 1772 the French chemist Antoine Lavoisier identified sulfur as an element, but not until 1809 did the French chemists Louis-Joseph Gay-Lussac and Louis-Jacques Therard prove it to be an element. It is easily recognized because of its yellow color and its penetrating odor when heated. It is found in elemental form, in sulfides and in sulfates. Large deposits are found in Louisiana and Texas. In ancient times it was used as a fumigant, in medicine, and in metallurgy. It has many current industrial uses, such as in acids, textiles, dyes, rocket fuels, and insecticides, etc. Chemical symbol is S.

Additional reading: *Van Nostrand's scientific encyclopedia;* p. 2708–2711.

Sulfuric Acid
CHEMISTRY

Sulfuric acid, a clear, oily liquid, is probably the most-used acid in industry. Uses include fertilizing, iron/steel production, pigments, and production of other chemicals. It was described around 1300 in the writings of a thirteenth century alchemist known as False Geber or Pseudo Geber, possibly a German. In 1595 the German alchemist Andreas Libau gave detailed descriptions of how to produce this acid. Modern production uses the contact process, which gives very pure acid of any desired strength.

Additional reading: *Illustrated science and invention encyclopedia;* v. 1, p. 13–14.

Sundial
HOROLOGY

For thousands of years Egyptians and others in the Middle East measured time by the position of a shadow of an upright stick in the ground. The stick or post was known as the gnomon. The hour was probably devised by the Egyptians around 4000 B.C. Because of the seasonal variation in the length of the day, the hour lines on a sundial would have to be slightly curved, with markings for the season of the year. This improved version was invented by Berosos (Babylonia) around 260 B.C.

Additional reading: *Illustrated science and invention encyclopedia;* v. 12, p. 2294–2295; v. 20, section 17.

Sunspot
ASTRONOMY

Sunspots, which are dark areas visible on the sun's surface, have been viewed and recorded for centuries. Records from China, for example, date back to around 100 B.C. Galileo Galilei, the Italian astronomer, was the first scientist to use a **telescope**. He recognized sunspots around 1610. These spots vary in size, in location, and in duration; there is evidence that they reach a peak in occurrence every 11 years. As a way to standardize measurements, observers use a formula called the Zurich Sunspot Number, devised in 1849 by the Swiss astronomer Rudolf Wolf. Normal ground observations have been supplemented by **space probes** and **X-ray astronomy**.

Additional reading: Buedeler. *The fascinating universe;* p. 165–176.

Supercharger
MECHANICAL ENGINEERING

In order to increase the power of an internal combustion engine, it is necessary to add more air to the cylinders. A supercharger acts as an air pump and compressor to accomplish this. The first rotary air blower was designed in 1866 by J. D. Roots (U.S.). In 1926 a supercharger was designed by Dugald Clerk (English). Around the mid-1940s the turbocharger was developed; it recycles exhaust gases back through the supercharger, rather than wasting the energy of those gases. The development of jet aircraft gave impetus to the improvement of the turbocharger.

Additional reading: *Illustrated science and invention encyclopedia;* v. 17, p. 2296–2297.

Supercomputer
COMPUTER SCIENCE

In 1976 the first commercial model of a supercomputer was created, designed by Seymour Cray (U.S.) Containing 200,000 integrated circuits, it could carry out 150 million operations per second. It was followed by such computers as Control Data's Cyber, which could perform 700 million operations per second. Speedier models are being invented regularly by computer companies. One example is the parallel processor, which refers to a computer which can work on more than one problem at a time. They began to appear around 1970. The ability to handle several tasks at once means parallel processors work entirely differently than conventional computers, even the very large, fast supercomputers, which have the traditional single-processor design and whose speed is in their size.

Additional reading: Giscard d'Estaing. *World almanac book of inventions;* p. 298.

Superconductivity
CRYOGENICS

When metals are cooled to temperatures near absolute zero, most of them become excellent conductors of electricity, with **electrons** flowing without significant resistance. This was discovered in 1911 by Kamerlingh Onnes (Holland). Not until around 1951 was the process understood due to the work of the American scientists John Bardeen, Leon Cooper, and John Schrieffer. In 1961 some scientists at Bell Laboratories discovered a group of superconductors which can carry high currents in very high magnetic fields. One major application has been the generation of very high magnetic fields for the bubble chambers used in nuclear physics.

Additional reading: *Illustrated science and invention encyclopedia;* v. 17, p. 2300–2302.

Supernova
ASTRONOMY

Aging stars which have a mass greater than about 1.4 times that of the sun (the Chandrasekhar Mass, discovered in 1939 by the Indian astronomer Subrahmaryan Chandrasekhar) will never become **White Dwarf stars.** But they may suddenly explode, following a last few weeks in which their original luminosity increases about one billion times. Three were known in the past ten centuries (1054, 1572, and 1604). The one in 1572 was observed by the Danish astronomer Tycho Brahe, whose name for it (Nova Stella, or New Star) gave such exploding stars their present name. The one in 1604 was observed by the German astronomer Johannes Kepler; in the period 1934-1938 the Swiss astronomer Fritz Zwicky observed 12 of them. The most recent one was observed in 1987 by Ion Shelton (Canada).

Additional reading: Buedeler. *The fascinating universe;* p. 222–224.

Supersonic Flight
AEROSPACE ENGINEERING

Many years of **aircraft** design passed before airplanes were close to flying at or above the speed of sound (around 750 miles per hour at 32 degrees Fahrenheit). Tests in **wind tunnels** resulted in better streamlining. The first supersonic flights occurred in 1947 when the American pilot Charles Yeager flew the X-1 rocket plane. Supersonic commercial flights did not commence until 1970, when the British-French Concorde plane began regular service, crossing the Atlantic in three hours. Supersonic flight is measured in terms of Mach numbers, named in honor of the Czech-born physicist Ernst Mach, who did pioneering work in 1887 on the effects on an object upon reaching the speed of sound.

Additional reading: Asimov. *Asimov's new guide to science;* p. 441–442.

Surgical Glove. *See* Antisepsis.

Surveying
CIVIL ENGINEERING

Surveying was known as far back as 3600 B.C. in Babylonia; Egyptian pyramids built around 2900 B.C. used surveyors. Plane surveying assumes the areas surveyed are on a horizontal plane, while geodetic surveying takes into account the curvature of the earth. In 1527 a theodolite for measuring angles in surveying was invented by Martin Weldseemuller (Germany). In 1533 the practice of triangulation surveying began, introduced by Gemma Frisius (Holland). In recent years gas **lasers** have been used to make a geodometer, which can measure distances accurately up to 15 miles. For distance up to 40 miles, a tellurometer (which uses a radar beam) is appropriate.

Additional reading: *Illustrated science and invention encyclopedia;* v. 17, p. 2309–2312.

Suspension Bridge CIVIL ENGINEERING

The very earliest suspension bridges were bunches of vines or creepers alongside streams, tied to tree trunks. One of the first professionally built suspension bridges was made in 1800 by John Finley (U.S.). He used iron chains from which hung vertical rods holding the road bed. Around 1820 suspension bridges were built across some Scottish rivers using wrought iron chains designed by Sir Samuel Brown (England). The first to use wire cables was built by Marc Seguin (France) in 1825. The famed Brooklyn Bridge, designed by John Roebling, and completed by his son W. A. Roebling in 1883, was the first to use cables which had a wrapping of wire around parallel wires rather than twisted wires. The Verrazano Bridge (1964) carries 12 lanes of traffic, spanning over 4,200 feet.

Additional reading: *Illustrated science and invention encyclopedia;* v. 3, p. 353–361.

Suture MEDICINE

Alexis Carrel, a French-born American surgeon, developed sutures in 1902. His technique allowed him to delicately close a **blood vessel** cut during a surgical incision. Carrel was awarded the 1912 Nobel Prize in medicine and physiology for his work on vascular sutures.

Additional reading: Sourkes. *Nobel prize winners in medicine and physiology 1901-1965;* p. 74–78.

Sweeper. *See* **Carpet Sweeper.**

Synapse ANATOMY

Heinrich Wilhelm Gottfried von Waldeyer-Hartz, a German anatomist, was the first person to observe that cells found in the nervous system never actually met. His 1891 observation was soon thereafter confirmed by Italian histologist Camillo Golgi. Applying his method of cellular staining with silver salts, Golgi verified that a gap between the cells did exist. In 1897 an English neurologist, Charles Scott Sherrington, coined the word "synapse" to describe this gap between the cells.

Additional reading: Asimov. *Asimov's new guide to science;* p. 816–833.

Synchrocyclotron. *See* **Cyclotron; Synchrotron.**

Synchrotron NUCLEONICS

During the years 1945 to 1960 several particle accelerators included the word "synchrotron" in their names. First was the synchrocyclotron built in 1946 to improve the **cyclotron.** It was designed by Edwin McMillan (U.S.). The proton synchrotron appeared in 1947 and the bevatron in 1954. In 1952 a proton synchrotron was built which was called a cosmotron because it produced particles in the energy range of cosmic waves. The most powerful synchrotron was named the strong-focusing synchrotron; it focused particles so they did not strike the walls of the channels in which they traveled.

Additional reading: Asimov. *Asimov's new guide to science;* p. 338–340.

Synthesizer (Music) ENGINEERING ACOUSTICS

The usual meaning of synthesizer is a device for creating sound electronically, principally musical sounds. The first one is credited to Robert Moog (U.S.) in 1967. Control is maintained over volume, tone quality, pitch, and rhythmic patterns. Computer memories can be used for repeating pre-set patterns as desired. Since then **microcomputers** have been built into synthesizers, and very elaborate, expensive models for professional musicians are available.

Additional reading: *Illustrated science and invention encyclopedia;* v. 17, p. 2327–2324.

Synthetic Diamond MATERIALS

Many efforts have been made to create synthetic diamonds, but it was not until 1955 that the first successful synthetic diamonds were prepared by the General Electric Company (U.S.). They are made in grit size (about 0.1 mm) and are in great demand in industry for grinding wheels and for sharpening carbide tools. They are made from carbon at high temperature and pressure.

Additional reading: *McGraw-Hill encyclopedia of science & technology;* v. V, p. 166.

Synthetic Gasoline PETROLEUM ENGINEERING

During World War II the shortage of gasoline encouraged the improvement of methods of producing synthetic gasoline from coal. Such a method, known as the Fischer-Tropsch process, had been developed in 1925, with the first commercial plant built in Germany in 1927. It consists of a mixture of coke and steam to produce carbon monoxide and hydrogen. The gases are then put into contact with a catalyst to form a mixture of hydrocarbons. In the alternative hydrogenation method, crushed coal is mixed with a heavy oil (derived from a later stage of the process) along with a catalyst such as tin oxalate. The slurry is then heated with hydrogen at high pressure. Both methods use fractional distillation to separate the different products, including gasoline.

Additional reading: *Illustrated science and invention encyclopedia;* v. 13, p. 1720–1721.

Syringe MEDICINE

Evidence has been found that ancient Romans used a nasal syringe made of metal. A small suction syringe used to clean wounds was invented in the early 1700s by French surgeon Dominique Anel. In 1853 Alexander Wood (Scotland) invented a hypodermic syringe, making the use of intravenous **anesthesia** possible.

Additional reading: Bennion. *Antique medical instruments;* p. 169–176.

T

Table of Elements. *See* **Periodic Table of Elements.**

Tabulating Machine. *See* **Calculating Machine; Punched Card.**

Tachometer MECHANICAL ENGINEERING

A device that measures the number of revolutions per minute of a rotating body is called a tachometer. The first one is said to have been constructed around 1840 by Bryan Donkin (England) for use in a mill. The greatest use now is for automobiles, particularly racing cars and large trucks. Several types currently exist, improving on the original mechanical drives. Most recent models operate electronically, although diesel engines (not having an electric ignition system) use magnetic detection of rotating ferrous pieces.

Additional reading: *Illustrated science and invention encyclopedia;* v. 17; p. 2330–2331.

Tank (Military) ORDNANCE

Early models of self-propelled armored vehicles known as tanks date back to 1908 when an English inventor named David Roberts developed a crude model; then in 1912 the Austrian inventor Gunter Burstyn demonstrated a cannon-mounted tank. Neither attracted any attention. French and British studies of tanks began around 1913, and in 1915 the British built a prototype, seeing battle action in 1916. After that, many countries began to make them, with speed improving from three miles per hour around 1918 to more than thirty miles per hour by the

1950s. Many changes have been made, including some light enough to be carried by planes. Amphibious models have also been created.

Additional reading: *Illustrated science and invention encyclopedia;* v. 17, p. 2337–2338.

Tantalum CHEMISTRY

A strong, ductile, corrosion-resistant metal, tantalum, atomic number 73, was discovered in 1802 by the Swedish chemist Anders Ekeberg, although for the next 40 years there were conflicting studies about the separate identities of columbium and tantalum. It is almost always found in minerals which also contain niobium. Much of it is obtained as a by-product in the smelting of tin. Its greatest use is in electrical capacitors and for corrosion-resistant materials, as well as alloys for turbine parts. Chemical symbol is Ta.

Additional reading: *Van Nostrand's scientific encyclopedia;* p. 2754–2755.

Tape. *See* **Adhesive; Magnetic Recording.**

Tape Recorder. *See* **Magnetic Recording; Videocassette Recorder.**

Tauon NUCLEAR PHYSICS

An accelerator that bombarded high-energy **positrons** with high-energy **electrons** enabled the American physicist Martin Perl in 1974 to detect a superheavy electron. It was named a tau electron, a term frequently shortened to tauon. It has a mass about 3,500 times as heavy as an electron, lasting only a trillionth of a second before breaking down to a muon.

Additional reading: Asimov. *Asimov's new guide to science;* p. 356.

Technetium CHEMISTRY

Technetium, atomic number 43, does not occur in nature but is obtained by bombardment of molybdenum with deuterons. All its isotopes are radioactive. It was discovered by the French physicist Carlo Perrier and the Italian-born physicist Emilio Segre (U.S.) in 1937. It is produced by processing of nuclear fuel waste solutions. Its main use so far has been as a tracer in medical treatment and as a superconductor. Chemical symbol is Tc.

Additional reading: *Van Nostrand's scientific encyclopedia;* p. 2763.

Teeth DENTISTRY

In 1778 John Hunter (England) published "A Practical Treatise on the Diseases of the Teeth," and classified teeth into molars, cuspids, and incisors. He also established that tooth decay began on the tooth's surface.

Additional reading: Ackerknecht. *A short history of medicine;* p. 134–135.

Telegraph COMMUNICATIONS

It had been known since the middle of the eighteenth century that the passage of static electricity along a wire could be observed at the end of the wire. This led to early attempts at telegraphy, such as a different wire for each letter of the alphabet! In 1837 an experimental telegraph system was installed in England by the inventors Charles Wheatstone and William F. Cooke; it used needles which pointed to a letter. In the United States the inventor Samuel Morse benefited in his work from what he could learn from the American physicist Joseph Henry, who had in effect created a telegraph system in 1831, but had never patented it. Morse got a patent in 1837, and sent a message in 1844 between Baltimore and Washington on the first telegraph system. (*See also* Submarine Cable)

Additional reading: Feldman. *Scientists & inventors;* p. 122–123.

Telephone COMMUNICATIONS

The patent of Alexander Graham Bell (U.S.), granted him in 1876, was decided by the U. S. Supreme Court to win out over that of Elisha Gray, who filed for a patent just two hours after Bell. In 1875, an accidental incident led Bell to invent a telephone rather than trying to improve a telegraph for Western Union. His telephone was improved by a circuit designed by Thomas Alva Edison (U.S.) in 1877. The first coin-operated telephone was patented in 1889 by William Gray (U.S.). A manual exchange was opened in 1878. The first automatic exchange was patented by Almon Strowger (U.S.) in 1901. The first dial phone was patented in 1923 by Antoine Barnay (France). The first electronic exchange opened in 1965. (*See also* Communications Satellite; Submarine Cable)

Additional reading: *Illustrated science and invention encyclopedia;* v. 17, p. 2356–2362.

Teleprinter COMMUNICATIONS

The teleprinter is a communication device that sends coded signals recorded on paper tapes which have been input in a typewriter-like machine. Its speed and ease of operation made it popular. The first one was invented in 1906 by Jay Morton (U.S.) and Charles Crum (U.S.), although a crude version had been invented in the 1890s by the English telegrapher Frederick Creed. By the 1920s some 30,000 offices had teleprinters. As their popularity grew, a network of machines was established, known as the Telex system, having around a million subscribers. A special **communications code** was agreed upon internationally.

Additional reading: *Illustrated science and invention encyclopedia;* v. 17, p. 2366–2368.

Telescope ASTRONOMY

The first known telescope was invented in 1608 by a Dutch optician named Hans Lippershey. Even though the Dutch tried to keep it a secret, Galileo Galilei, the noted Italian astronomer, heard about it and by 1610 had improved it for use in astronomy. Because lenses caused a dispersion of light into obscuring rings of colors, Sir Isaac Newton (England) in 1668 built a reflecting telescope in which a parabolic mirror magnified an image. In 1757 the English optician John Dolland prepared lenses made of two kinds of glass so that the bothersome spectra were not created; this was an achromatic lens. In 1897 a large 40-inch lens was built in this way. Many larger telescopes have since been built, and **radio astronomy** has created new telescopes.

Additional reading: Asimov. *Asimov's new guide to science;* p. 57–58, 651.

Teletypewriter. *See* **Teleprinter.**

Television COMMUNICATIONS

The earliest form of television was invented in 1926 by the Scottish inventor John Baird, who used a mechanical **scanning disk**. Although in 1928 he was able to send a transmission from London to New York, the limitations on his design made major improvements impossible. The Russian-born American inventor Vladimir Zworykin, patented the iconoscope, the first television scanning camera, in 1923. In 1924 he patented the kinescope, the television picture tube. Both inventions rely on streams of electrons either for scanning (in the camera) or for creating images on a fluorescent screen (in the receiver). His first practical model was demonstrated in 1938. Widespread ownership of television sets did not occur until the 1950s. (*See also* Color Television)

Additional reading: Feldman. *Scientists & inventors;* p. 276–277.

Telex. *See* **Teleprinter.**

Tellurium CHEMISTRY

Tellurium is a silver-white brittle semi-metal, atomic number 52, that is the only element forming several minerals with **gold**. It was first identified by Franz Muller (Baron de Reichenstein), a mining engineer from Romania, in 1784, while studying a bluish-white ore of gold from Transylvania. Ores containing tellurium are found in California, Colorado, Ontario, and Mexico. Most of it is produced as a by-product of refining **copper** and **lead.** It is very toxic. The main use is in alloys and in some electronic components, such as solar cells and infrared detectors. Chemical symbol is Te.

Additional reading: *Van Nostrand's scientific encyclopedia;* p. 2789.

Tellurometer. *See* **Surveying.**

Terbium CHEMISTRY

A silvery-colored metal, terbium, atomic number 65, was first identified in 1843 by Carl Mosander, a Swedish mineralogist. It occurs in apatite and xenotime. Terbium is produced as a by-product in the processing of yttrium. Its uses are few, but include image intensifying for **X-ray** screens, **color television** tubes, and **semiconductors**. Chemical symbol is Tb.

Additional reading: *Van Nostrand's scientific encyclopedia;* p. 2793–2794.

Terramycin. *See* **Antibiotic.**

Terrestrial Magnetism GEOPHYSICS

The first scientist known to have stated that the earth has two magnetic poles was the English physician and physicist William Gilbert, who wrote about magnetism in 1600. He also pointed out the fallacies then prevalent about many other aspects of magnetism. He also wrote on static electricity. The Irish astronomer Sir Edward Sabine in 1852 was the first to correlate disturbances in the earth's magnetic field with **sunspots.** He also set up a global network of stations to measure magnetic data, which led to a large compilation of information. The evidence that the earth's magnetic poles have undergone **magnetic reversal** was found in a study in 1960.

Additional reading: Asimov. *Asimov's biographical encyclopedia;* p. 90–91, 308–309.

Test Tube Baby. *See* **Artificial Insemination.**

Tetanus Vaccine MICROBIOLOGY

The germ that caused tetanus or lockjaw, tetanus bacillus, was discovered in 1884 by Arthur Nicolaier (Germany). In 1890 Emil Adolf von Behring (Germany) discovered that he could produce an immunity to tetanus in animals by injecting them with graded doses of blood serum from other animals afflicted with tetanus. A portion of the serum from the immunized animals could be used to render other animals immune to the disease. Von Behring gave the name "antitoxin" to the serum used for injections. Von Behring worked with Baron Shibasaburo Kitasato, a Japanese bacteriologist, to establish the specific antitoxin for tetanus.

Additional reading: Asimov. *Asimov's new guide to science;* p. 679–683.

Tetracycline. *See* **Antibiotic.**

Tevatron NUCLEONICS

One of the modern particle accelerators is the Tevatron, built in 1982 at the Fermi National Accelerator Laboratory. One of its unique features provides for two streams of particles, circling in opposite directions until caused to collide head on with each other. This should provide four times as much energy as having particles strike a stationary target. Larger and larger projects are being planned, particularly by the United States and the Soviet Union.

Additional reading: Asimov. *Asimov's new guide to science;* p. 340–341.

Thallium CHEMISTRY

Thallium in metallic form, atomic number 81, is bluish-gray, changing to dark gray on standing. It was discovered in 1861 by Sir William Crookes, the English physicist, while searching for tellurium using spectroscopic methods. Thallium occurs in small amounts in such minerals as hematite and pyrite. It is produced by treatment of the **lead** and **zinc** concentrates. Its main use is in electronic encapsulation, in selenium rectifiers, and in low-temperature applications. Chemical symbol is Tl.

Additional reading: *Van Nostrand's scientific encyclopedia;* p. 2796–2797.

Theodolite. *See* **Surveying.**

Thermistor. *See* **Semiconductor.**

Thermocouple PHYSICS

A device for measuring temperatures, this consists of two metals which generate an electric current proportional to the temperature difference of the junctions of the metals. Based on the work of Thomas Seebeck (Germany) in 1821, thermocouple design did not make much progress until the 1950s, when research in new alloys led to better devices. Thermocouples now have many industrial and commercial uses now. (*See also* Seebeck Effect)

Additional reading: *Illustrated science and invention encyclopedia;* v. 18, p. 2405–2406.

Thermodynamics THERMODYNAMICS

Thermodynamics is a branch of physics which deals essentially with the conversion of energy from one form to another, including relationships between the properties of matter and the effects of change of temperature, pressure, etc. The work of Joule and Rumford (among others) in studying **heat** led to the first law of thermodynamics, sometimes referred to as the **conservation of energy** law. The second law was based on the work of Sadi Carnot (France), who theorized in 1824 on the ideal efficiency of a "perfect" heat engine. From this the German physicist Rudolf Clausius deduced in 1850 the second law, involving the concept of **entropy.** The third law, which was stated by Walther Nernst (Germany) in 1906, says the temperature of absolute zero cannot be reached.

Additional reading: *Illustrated science and invention encyclopedia;* v. 18, p. 2398–2402.

Thermoelectricity PHYSICS

This refers to three related effects which involve the conversion of electricity into heat and vice versa. The **Seebeck effect** involves creating an electric current from heat, the **Peltier effect,** concerns creating heat through electric currents, as does the **Thomson effect.** Credit for the first discovery of thermoelectricity goes to the German physicist Thomas Seebeck in 1821.

Additional reading: *Illustrated science and invention encyclopedia;* v. 18, p. 2403–2406.

Thermometer PHYSICS

Early in the seventeenth century the first thermometers appeared, using the expansion of air with temperature, forming a column of water along a narrow tube having an arbitrary scale. The early inventors included Galileo Galilei (Italy) in 1593. In 1654 Ferdinand II (Italy) developed the first sealed liquid-in-glass thermometer, with experiments made using either alcohol or mercury. Thermometers graduated between two fixed points were being built in the late seventeenth century, and Gabriel Daniel Fahrenheit (Poland) produced excellent thermometers of this type around 1717. He also developed the standard scale of temperatures that bears his name. Around 1860, Sir William Siemens (England) developed the electrical resistance thermometer. (*See also* Thermoscope)

Additional reading: *Illustrated science and invention encyclopedia;* v. 18. p. 2406–2408.

Thermonuclear Reaction. *See* **Nuclear Fusion; Thermonuclear Weapon.**

Thermonuclear Weapon NUCLEONICS

As powerful as the fission bomb (the atomic bomb) was in the 1940s, some scientists began to perceive that an even more deadly weapon would be a fusion bomb (the hydrogen bomb). It was opposed by the American physicist J. Robert Oppenheimer, but his views were ignored by President Truman in 1949. The project, however, was strenuously suppported by the American physicist Edward Teller around 1950, and work went ahead. The first hydrogen bomb was tested in 1952, followed soon by a Soviet hydrogen bomb. Other countries developed their bombs also. In the 1970s thermonuclear bombs were built which have a smaller blast effect but a larger amount of **nuclear fallout.** More damage would be done to people and less to property. (*See also* Nuclear Fission; Nuclear Weapon)

Additional reading: Asimov. *Asimov's new guide to science;* p. 474–476, 480–483, 484–488.

Thermos Bottle. *See* Dewar Flask.

Thermoscope — MEDICINE

In 1626 Italian physician Santorio Santorio, also known as Sanctorius, adapted the thermometer invented by Galileo Galilei (Italy) in 1593 for the purpose of measuring human temperature in a clinical setting. Santorio's device was called a "thermoscope."

Additional reading: Reiser. *Medicine and the reign of technology;* p. 110–121.

Thiamin. *See* Vitamin B.

Thomson Effect — PHYSICS

A part of the phenomena known as **thermoelectricity,** the Thomson effect states that heat flows into or out of a homogeneous conductor when an electric current is made to pass between two points of the conductor having different temperatures. It was discovered around 1850 by Lord Kelvin (William Thomson), a Scottish physicist.

Additional reading: *Illustrated science and invention encyclopedia;* p. 2404–2405.

Thorium — CHEMISTRY

Thorium, atomic number 90, is a soft, dark gray metal, all forms being radioactive. It was discovered in 1828 by the Swedish chemist Jons Berzelius. Its main source is monazite sand, found in Brazil, India, and North and South Carolina, from which it is produced by an acid treatment. Its main use is as a fuel for **nuclear reactors** and in certain alloys and metal production. Chemical symbol is Th.

Additional reading: *Van Nostrand's scientific encyclopedia;* p. 2814.

Thulium — CHEMISTRY

Thulium is a gray metal, atomic number 69, which was discovered in 1879 by Per Cleve, a Swedish chemist. It is the least abundant of the rare-earth elements. It occurs in apatite and xenotime, and is produced by a by-product in the treatment of yttrium. Its few uses include being a constituent in bubble devices and catalysts. Chemical symbol is Tm.

Additional reading: *Van Nostrand's scientific encyclopedia;* p. 2817.

Thymine. *See* Nucleic Acid.

Thyrotropin-Releasing Hormone. *See* TRH.

Thyroxine — BIOCHEMISTRY

In 1914 Edward Calvin Kendall (U.S.) isolated and named thyroxine, the active constituent of the thyroid **hormone** which is secreted by the thyroid gland.

Additional reading: Asimov. *Asimov's new guide to science;* p. 720–723.

Tidal Power — OCEANOGRAPHY

Harnessing the power of the tides has been considered for centuries. In the 1130s a mill was built near the mouth of the Rance River in France to use tidal power. Around 100 years later several tidal stations were located near Venice, and in 1582 a tide-driven pump was in use on the Thames River, built by Peter Morice (England). Since the 1960s a tidal plant on the Rance River has generated electricity successfully, using 25 turbogenerators, which operate in both directions of tidal flow.

Additional reading: Clark. *Works of man;* p, 341.

Tide — ASTRONOMY

Probably the first scholar to realize that the tides are caused by the influence of the moon was the Greek geographer Pytheas around 350 B.C. Around 200 B.C. the Greek astronomer Seleucus not only realized the influence of the moon but also the fact that the tides come at different times around the world. The calculation of the motions of ocean tides, due to the gravitational pull of the moon, was enhanced by the laws discovered around 1665 by the English scientist Sir Isaac Newton.

Additional reading: Downs. *Landmarks in science;* p. 147.

Time and Motion Study — INDUSTRIAL ENGINEERING

A motion study analyzes all the separate stages involved in a task, and a time study measures the amount of time each stage requires. The object of both studies is to evolve the most efficient way of performing a task and to establish standards for each step. In 1760 J. R. Peronet (France) made time studies in the manufacturing of pins. Around 1773 Matthew Boulton (England) and Richard Arkwright (England) studied factory efficiency. Robert Owen (Scotland) studied factory operations in the early nineteenth century, including inauguration of rest periods. In America, Frederick Taylor began time and motion studies around 1881; Lillian and Frank Gilbreth, beginning in 1911, emphasized motion studies, often filming tasks to study them more carefully.

Additional reading: *Illustrated science and invention encyclopedia;* v. 18, p. 2435–2437.

Tin — CHEMISTRY

Tin, atomic number 50, is a silver-white metal with a bluish tinge. Artifacts made of **bronze,** a tin-containing alloy, show that tin was known in Egypt as far back as 3500 B.C. There are no high-grade tin deposits, most tin being found in low-grade deposits in Bolivia. It is produced in large part by treatment of the mineral known as cassiterite. Tin is used extensively as an alloy, in coatings, and in compounds. Chemical symbol is Sn.

Additional reading: *Van Nostrand's scientific encyclopedia;* p. 2832–2834.

Tissue. *See* Histology.

Tissue Culture — CYTOLOGY

Ross G. Harrison, an American zoologist, invented tissue culture in 1907 while culturing fragments of the central nervous system of frogs. During the early 1900s, Alexis Carrel (U.S.) developed a technique that allowed him to keep small amounts of tissue alive in test tubes by nourishing them with blood circulated by a crude **artificial heart.**

Additional reading: Asimov. *Asimov's new guide to science;* p. 681–683.

Titanium — CHEMISTRY

A white brittle metal, titanium, atomic number 22, was discovered in 1790 by the English amateur chemist William Gregor. It is quite abundant in nature, occurring in almost all rocks and in many minerals. Rutile is the principal commercial interest, found in Australia and Florida. It is widely used in ship and airplane construction, military equipment and in industrial alloys. Chemical symbol is Ti.

Additional reading: *Van Nostrand's scientific encyclopedia;* p. 2835–2839.

TNT. *See* Explosive.

Tocopherol. *See* Vitamin E.

Toilet
BUILDING CONSTRUCTION

Sanitation was quite neglected for centuries, and it was not until 1596 that Sir John Harrington, a godson of Queen Elizabeth, devised a toilet, having a cistern as a source of water and a valve for releasing water. Although one was installed in the palace, the device was not widely used. In 1778 Joseph Bramah (England) patented a design, three years after a watchmaker, Alexander Cumming (England), had patented a valve type model. Both inventors' devices were installed in many homes, although the lack of sewers meant that cesspools and primitive outdoor facilities had to be used for many years.

Additional reading: *Illustrated science and invention encyclopedia;* v. 19, p. 2611–2613; v. 21, section 62.

Tokamak
PLASMA PHYSICS

One of the designs for containing the hot gases known as **plasma** has been a magnetic field known popularly as the tokamak (toroid camera with magnetic field). It was designed by the Soviet physicist Lee Artsimovich around 1959. Refinements of the device are under continuous study as it is a very complicated area of research.

Additional reading: Asimov. *Asimov's new guide to science;* p. 503.

Tool. *See* **Hand Tools; Machine Tool.**

Torque Converter. *See* **Automatic Transmission.**

Torsion Modulus. *See* **Modulus of Elasticity.**

Tracer
BIOCHEMISTRY

German biochemist Franz Knoop used the first tracer in biochemistry in 1904. Knoop demonstrated that he could feed labeled fat molecules to dogs and then trace the molecule to see what happened to it. He labeled the molecule by attaching a benzene ring to one end of the chain.

Additional reading: Asimov. *Asimov's new guide to science;* p. 583–386.

Track Tractor
AGRICULTURE

The first successful caterpillar type tractor was patented by Benjamin Holt (U.S.) in 1904. Commercially produced by 1908, the track tractor inspired the development of the military **tank.** (*See also* Internal Combustion Tractor)

Additional reading: Schlebecker. *Whereby we thrive;* p. 199–205.

Tractor. *See* **Internal Combustion Tractor; Track Tractor.**

Traffic Signal
MECHANICAL ENGINEERING

Special signals for controlling pedestrian and vehicular traffic date back to 1866 when a mechanical signal was invented by T. Hodgson (England); in 1868 it was installed near Westminster Abbey. With the advent of electricity early versions of the familiar lighted signals began to appear. In more recent years some use has been made of vehicle-actuated traffic lights, based on inductive metal loops buried in the roads; these proved to require less maintenance than rubber-covered pneumatic tubes, which wore out quickly. Still other techniques have been tried, such as using computers to sense traffic conditions and control the lights accordingly.

Additional reading: *Illustrated science and invention encyclopedia;* v. 18; p. 2462–2465.

Tranquilizer. *See* **Librium; Reserpine.**

Transducer
PHYSICS

A transducer is a device for converting energy from one form to another, such as a **microphone**, which converts sound energy into electrical energy. There are several types, such as inductive transducers, capacitive transducers, strain gages, ultrasonic transducers, and feedback transducers. Development of the various types dates back at least 100 years, when Thomas Alva Edison (U.S.) invented in 1877 a microphone in which sound waves impinging on carbon granules varied their electrical resistance; his invention was soon incorporated in early **telephones.** New types of transducers are regularly being invented. (*See also* names of specific types of transducers)

Additional reading: *Illustrated science and invention encyclopedia;* v. 18, p. 2470–2473.

Transfer-RNA
BIOCHEMISTRY

The American biochemist Mahlon Bush Hoagland showed the existence of small RNA molecules in **cytoplasm** during the 1950s. These molecules, which he called soluble-RNA, were eventually called transfer-RNA. English biochemist Francis Harry Compton Crick predicted the discovery of transfer-RNA in 1958. In 1964, transfer-RNA was completely analyzed by an American team headed by Robert William Holley (U.S.). Holley shared the 1968 Nobel Prize for medicine and physiology with Marshall Warren Nirenberg (U.S.) and Har Hobind Khorana (U.S.). (*See also* RNA; Messenger-RNA)

Additional reading: Gribbin. *In search of the double helix;* p. 258–294.

Transference
PSYCHOLOGY

Austrian psychiatrist Sigmund Freud developed his ideas on transference during the 1880s and 1890s. He used the term to describe a patient's involvement with the therapist during **psychoanalysis.** Freud believed transference was central to the therapeutic process.

Additional reading: Alexander. *The history of psychiatry;* p. 188–210.

Transforming Principle
GENETICS

Oswald T. Avery (U.S.), Colin Munro MacLeod (Canada), and Maclyn McCarty (U.S.), bacteriologists working on pneumococcal transformation at the Rockefeller Institute, presented in 1944 the first experimental evidence that **DNA** is the carrier of genetic information in all living organisms. Their data demonstrated that a **nucleic acid**, not a **nucleoprotein**, is the hereditary agent. Their research did not receive full acceptance until 1952 when work by Alfred Hershey (U.S.) and Martha Chase (U.S.) on phage particles showed that only DNA actually entered the bacterial host cell, thus demonstrating that the genes resided in DNA.

Additional reading: McCarty. *The transforming principle.*

Transistor
SOLID-STATE PHYSICS

The limitations of the vacuum tube hindered efforts to miniaturize electronic equipment. Following a careful study of the properties of **semiconductors,** three American physicists discovered that introducing impurities in semiconductors gave special advantages—a solid-state material that would not only rectify a current but also amplify it. Thus in 1948 William Shockley, Walter Brattain, and John Bardeen (all U.S.) invented the transistor, which within a short time made most vacuum tubes obsolete. Later it became possible to etch transistors on small pieces of silicon in **integrated circuits.**

Additional reading. Asimov. *Asimov's new guide to science;* p. 447–451.

Translocation
GENETICS

American biologist Barbara McClintock showed in 1934 that the nucleolus organizer of Zea mays could be split by a "translocation" and that the resulting pieces each were capable of organizing a separate nucleolus. She continued her research and during the 1940s demonstrated that a **chromosome** segment, a regulator gene, was able to transfer from its usual position to a new one in the same or in a different chromosome. McClintock's findings were ignored until the 1970s and early 1980s when other researchers began finding evidence of **gene** mobility. McClintock was awarded the 1983 Nobel Prize in medicine and physiology for her research.

Additional reading: Gribbin. *In search of the double helix;* p. 78–82, 295–320.

Transmission. *See* **Automobile Transmission.**

Transuranium Element
NUCLEAR PHYSICS

Elements heavier than **uranium** (atomic number 92) were sought for as early as 1934 when the Italian-born physicist Enrico Fermi (U.S.) bombarded uranium with **neutrons** and found it resulted in element 93 (**Neptunium**). Later it was realized he had accomplished this by splitting the uranium atom. The way to discover more elements was seen to be by bombardment of appropriate elements, which began in earnest in 1944, led by the American physicist Glenn T. Seaborg. Scientists are continuing to look for what have been called super-heavy elements. (*See also* Nuclear Fission)

Additional reading: Asimov. *Asimov's new guide to science;* p. 278–281.

TRH
BIOCHEMISTRY

Polish-born Andrew Victor Schally (U.S.), a biochemist, and French-born Roger Guillemin (U.S.), a physiologist, discovered in 1968 and 1969 that thyrotropin-releasing hormone (TRH), a substance produced by the hypothalamus, controls the pituitary gland. TRH is one among other peptides that have dual functions as both **hormones** and as neurotransmitters. Schally and Guillemin shared the 1977 Nobel Prize in physiology and medicine for their work.

Additional reading: Crapo. *Hormones: the messengers of life;* p. 11, 171–172.

Trichloromethane. *See* **Chloroform.**

Trigonometry
MATHEMATICS

Trigonometry, the study of triangles, began with the work of the Greek scholar Hipparchus around 140 B.C. He laid down principles for the measurement of triangles and worked out accurate tables of the ratios of the sides of a right triangle, thus giving him the credit for being the founder of trigonometry. His work was extended by another Greek scholar, Ptolemy of Alexandria, around 150 A.D. Using trigonometry, Ptolemy was able to estimate the distance of the moon from the earth. Trigonometric abbreviations such as sin, cos, tan, etc., were first used around 1615 by the Dutch mathematician Albert Girard.

Additional reading: Asimov. *Asimov's biographical encyclopedia;* p. 33–34, 42–43.

Trophic-Dynamic Concept
ECOLOGY

In 1942, R. L. Lindeman (U.S.) developed the trophic-dynamic concept, which describes energy flow through the **ecosystem.** Lindeman's concept signalled the beginning of modern ecology.

Additional reading: McIntosh. *The background of ecology;* p. 196–197.

Tropopause. *See* **Stratosphere.**

Tryptophan
BIOCHEMISTRY

Tryptophan, an **amino acid** building block of **protein,** was discovered in 1900 by English biochemist Sir Frederick Gowland Hopkins. Hopkins found that tryptophan was not manufactured in the body and, therefore, had to be supplied through diet. Hopkins introduced the concept of "essential amino acids."

Additional reading: Asimov. *Asimov's new guide to science;* p. 699–701.

Tsunami
OCEANOGRAPHY

Wide shallow waves which are generated by **earthquakes** or underwater volcano activity are called tsunamis. They may be up to 100 miles long and travel as fast as 450 miles an hour. They have caused great destruction along coastal areas and are quite difficult to predict. One earth scientist who made detailed studies of the theory of tsunamis in the 1930s was Katsutada Sezawa (Japan).

Additional reading: Fairbridge. *Encyclopedia of oceanography;* p. 941–943.

Tuberculin Test
IMMUNOLOGY

Robert Koch, a German bacteriologist, discovered in 1882 that tuberculosis was caused by the tubercle bacilli. While attempting to find a cure for tuberculosis, he developed a tuberculin test in 1890 that proved effective in determining sensitization to the bacilli. Although Koch did not discover a cure, he did find a valuable diagnostic tool, and was awarded the 1905 Nobel Prize in medicine and physiology for his contributions in connection with tuberculosis.

Additional reading: Bettmann. *A pictorial history of medicine;* p. 288–289.

Tungsten
CHEMISTRY

Tungsten, a silver-white to steel-gray metal, atomic number 74, was mistakenly described in 1574 by Lazarus Ecker as a mineral of tin when he examined wolframite. In 1781 the Swedish chemist C. W. Scheele studied a mineral called tungsten and considered it also to be a tin ore. It was not until 1783 that two Spanish brothers, J. J. and F. de Elhuyar, produced the first metallic tungsten and also showed that tungsten was a constituent of wolframite, called wolfram outside the United States. Tungsten is found in pegmatites and batholiths. Deposits are chiefly found in the United States, China, and Korea. Much of the amount produced is used in making cutting tools and wear-resistant parts; other uses include wires for light bulbs and industrial alloys. Chemical symbol is W.

Additional reading: *Van Nostrand's scientific encyclopedia;* p. 2869–2871.

Tunnel
CIVIL ENGINEERING

Over the centuries tunnels have served many purposes, whether for roads, railroads, or water supplies. One of the oldest was supposedly built for Queen Semiramis of Babylonia around 2170 B.C.; it was located under the Euphrates River and was 600 feet long. A famous tunnel built in Italy for Emperor Claudius around 50 A.D. was a 3.5 mile structure for draining a lake; portions found 1,800 years later were still intact. A difficult tunnel was built under the Thames River by the French engineer Marc Brunel in 1843. He used innovative techniques such as reinforced Portland cement and a large metal shield covering workers doing excavating. Other notable tunnels built beginning in the 1850s went through the Alps.

Additional reading: Clark. *Works of man;* p. 261–276.

Tunnel Diode. *See* **Semiconductor.**

Turbine. *See* **Gas Turbine; Hydroelectric Station; Jet Engine; Steam Turbine.**

Turbocharger. *See* **Supercharger.**

Turbojet Engine. *See* **Jet Engine.**

Turboprop Engine. *See* **Jet Engine.**

Twine Binder
AGRICULTURE

The twine binder was invented and patented in 1874 by John Appleby (U.S.) and was used to bundle and tie sheaves of grain in conjunction with the work of the **reaper.** Appleby's invention made the use of twine practical and an improvement over using wire which animals easily ingested with fatal results. By 1882 the twine binder completely replaced the use of wire.

Additional reading: Schlebecker. *Whereby we thrive;* p. 190–191.

2, 4-D. *See* **Herbicide.**

TWX. *See* **Teleprinter.**

Typesetting
GRAPHIC ARTS

Although improvements were slowly being made in the nineteenth century in printing presses, the setting up of type was essentially a manual job done by hand. Finally in 1884 the German inventor Ottomar Mergenthaler devised a machine to set type automatically, directed by an operator at a keyboard. He named it the Linotype machine. A rival device, the Monotype machine, was built in 1887 by the American inventor Tolbert Lanston. Both could set at least 5,000 pieces of type an hour. Not until the 1950s were they superseded by more modern machines such as the device for **phototypesetting.**

Additional reading: Feldman. *Scientists & inventors;* p. 215–217.

Typewriter
GRAPHIC ARTS

The first known patent for a typewriter was issued in 1714 to an Englishman named Henry Mills, but little is known about it. Other models were produced, but for practical purposes the first commercial machine was not invented until 1867 when the American printer Christopher Sholes and the American inventor Carlos Glidden built a working machine that they patented the next year. This model was put on the market in 1874 by Remington & Sons. Many improvements have been made since then, including replacement of mechanical models with electric and electronic models. The American inventor R. G. Thomson developed an electric typewriter in 1933, while an American firm produced the first electric typewriter with a memory in 1965. (*See also* Word Processing)

Additional reading: *Illustrated science and invention encyclopedia;* v. 18, p. 2504–2508.

U

Ultracentrifuge
CHEMISTRY

Swedish chemist Theodor H.E. Svedberg developed the ultracentrifuge in 1923 for the purpose of producing a very strong centrifugal force that would move tiny colloidal particles out of solution. In particular, Svedberg wanted to separate out **protein** molecules and needed forces stronger than those produced by the **centrifuge.** He was awarded the 1926 Nobel Prize for chemistry as a result of his contribution.

Additional reading: Asimov. *Asimov's new guide to science;* p. 556.

Ultrasonics
ACOUSTICS

Sound waves having a frequency above the upper audible range for humans (around 20,000 vibrations per second) are called ultrasonic sound. This topic was studied in 1880 by the French chemists Pierre Curie and his brother Jacques, who discovered that the **piezoelectric effect** could serve as a means of creating electrical impulses through mechanical pressure on certain crystals. Using high frequencies, they and several others began to investigate such waves as a means of detecting submarines. Since then there have been many applications of ultrasonics, including the testing of metals for defects, ultrasonic surgical tools, and underwater detection. (*See also* Sonar; Transducer)

Additional reading: *Illustrated science and invention encyclopedia;* v. 19, p. 2515–2518.

Ultraviolet Radiation
PHYSICS

The study of the spectrum of sunlight at the beginning of the nineteenth century revealed the existence of nonvisible waves. The German physicist Johann Ritter found in 1801 that silver nitrate, which breaks down to metallic silver when exposed to blue or violet light, broke down even more when it was exposed to incoming "light" just beyond the violet portion of the spectrum. Thus he proved that sunlight contains more than just the visible wavelengths, just as the discoverer of **infrared radiation** had concluded the year before.

Additional reading: Asimov. *Asimov's new guide to science;* p. 60–61.

Uncertainty Principle
QUANTUM MECHANICS

The problem of understanding the nature of **subatomic particles** was studied by the German physicist Werner Heisenberg. In 1927 he startled the world with the statement that it was impossible to measure simultaneously both the position as well as the momentum (mass times velocity) of a body, his uncertainty principle. It is primarily in connection with measuring subatomic particles that this is important. For example, using an **electron microscope** to study an atom might dislodge **electrons** so that the process of measurement changed the object being studied. He also developed a system called matrix mechanics, which was found to be less preferable than **wave mechanics.**

Additional reading: Asimov. *Asimov's new guide to science;* p. 407–410.

Unconscious
PSYCHOLOGY

Socrates, a Greek philosopher who lived from 469 to 399 B.C., was the first Western thinker to assert that mental activities occurred outside the realm of awareness and in the unconscious. Plato, another Greek philosopher living between 428 and 347 B.C., discussed the existence of unconscious motivation. In 1895 Josef Breuer (Austria) and Sigmund Freud (Austria) suggested that at the root of most emotional disorders could be found unconscious traumatic memories. Freud's development of **psychoanalysis** as a method of treatment was greatly influenced by ideas on the unconscious.

Additional reading: Murray. *A history of Western psychology;* p. 298–299.

Underwater Cable. *See* Submarine Cable.

Unified Field Theory
PHYSICS

The need for finding a single field which would apply to both electromagnetic and gravitational fields was important to Albert Einstein (U.S.). Around 1915 he began a fruitless search that was to occupy half his life. The Americans Steven Weinberg and Sheldon Glashow and the Indian physicist Abdus Salam took up the study in the late 1960s. In 1979 the three shared the 1979 Nobel prize for physics. By then two more short-range fields had been discovered. In the 1970s, the American Alan E. Guth used the unified field in his study of the origin of the universe.

Additional reading: Asimov. *Asimov's new guide to science;* p. 370–372.

Universal Joint
MECHANICAL ENGINEERING

A universal joint is a linkage that transmits rotation between two shafts in the same plane but not in a straight line with each other. The first ones were proposed in 1551 by the Italian mathematician Girolamo Cardano and are often still called Cardano joints. The first practical model was invented around 1678 by the English physicist Robert Hooke, and are called Hooke joints. Other variations have been created; one, called a constant velocity joint, is used regularly in front wheel drive automobiles.

Additional reading: *Illustrated science and invention encyclopedia;* v. 19, p. 2527–2528.

Universe's Origin
ASTRONOMY

The origin of the universe has attracted the interest of many scientists over the centuries. In 1785 the Scottish naturalist James Hutton published a work stating that the earth must be millions of years old, rather than the thousands of years on which the Biblical version of creation was based. Later use of radioactivity analyses proved that the world had to be over 4 billion years old. In the 1920s scientists began to propose that the universe was formed as a result of a violent explosion, known informally as the "Big Bang" theory. This was popularized by the Russian-born scientist George Gamow in 1948, based on the work in 1927 of the Belgian astronomer Georges Lemaitre and some earlier scientists. (*See also* Earth's Formation; Solar System's Origin)

Additional reading: Asimov. *Asimov's new guide to science;* p. 37–45.

Uranium CHEMISTRY

Uranium, atomic number 92, was discovered by the German chemist Martin Klaproth in 1789 while working with pitchblende. In 1896 the French chemist Henri Becquerel discovered its **radioactivity**. Uranium was the first element to undergo fission, when it was bombarded with **neutrons** in 1934 by the Italian-born physicist Enrico Fermi (U.S.). Uranium deposits are found throughout the world, with Canada, the United States, France, and East Germany as leading sources. Its main use is as a fuel for **nuclear reactors**. Chemical symbol is U.

Additional reading: *Van Nostrand's scientific encyclopedia;* p. 2893–2898.

Uranus ASTRONOMY

The planet Uranus is the dimmest planet, barely visible to the naked eye. It was not until 1781 that the German-born astronomer Sir William Herschel (England) observed an object that was disk-shaped and not hazy like a comet; it was the first new planet to be discovered in historic times. The calculation of its orbit later that year by the Swedish astronomer Anders Lexell confirmed it was a planet. In 1787 Herschel discovered two satellites of Uranus, and two more were observed in 1851 by the English astronomer William Lasell. In 1948 the Dutch-born astronomer Gerard Kuiper (U.S.) discovered a fifth satellite. In 1977 the American astronomer James Elliot observed thin rings around Uranus. The Voyager space probe observed Uranus in detail in 1986.

Additional reading: Asimov. *Asimov's new guide to science;* p. 140–142.

Urea. *See* **Organic Compound Synthesis.**

V

Vacuum Cleaner MECHANICAL ENGINEERING

John J. Thurman (U.S.) invented the first powered vacuum cleaner in 1899. A patent for a vacuum cleaner was issued in 1901 to Herbert Cecil Booth (England). It used a five-horsepower piston engine for its operation. The earliest vacuum cleaners were mounted on horse-drawn carriages and ran hoses into buildings needing to be cleaned. In 1907 the first electric vacuum cleaner was devised by J. M. Spangler (U.S.). In the mid-1920s portable models became available.

Additional reading: *Illustrated science and invention encyclopedia;* v. 19, p. 2535–2536.

Valence. CHEMISTRY

In the 1850s chemists began to note that atoms of certain elements could combine with only a fixed number of atoms of other elements. In 1852 the English chemist Sir Edward Franklin stated this theory; the phenomenon was later called valence, a measure of the combining power of the elements. His discoveries influenced the work of others, such as Faraday's laws of **electrolysis,** or the rise and fall of valence in the arrangement of elements in Mendeleev's **periodic table of elements.**

Additional reading: Asimov. *Asimov's biographical encyclopedia;* p. 434.

Valproic Acid PHARMACOLOGY

In 1963 Pierre Eymard (France) discovered the anticonvulsant properties of valproic acid, also known as dipropylacetic acid. In 1967 it was marketed under the trade name Epilim, and was used to control epileptic seizures.

Additional reading: Sneader. *Drug discovery;* p. 46–47.

Van Allen Belt ASTROPHYSICS

Several belts of trapped radiation were discovered in the earth's atmosphere in 1958 by James Van Allen (U.S.) and his colleagues by means of radiation detectors carried on **satellites** in the Explorer series. The belts were found to be some 1,500 miles from the earth, extending some 40,000 miles in the direction of the sun and extending outward for a million miles or so. The belts, now called the magnetosphere, consist of charged particles, essentially **electrons** and **protons.**

Additional reading: Asimov. *Asimov's new guide to science;* p. 232–233.

Van de Graaf Generator NUCLEONICS

A device that separated **electrons** from **protons** by means of a moving belt was devised by the American physicist Robert Van de Graaf in 1931. Named the Van de Graaf generator, it was a big improvement over previous devices for developing energy for bombarding particles.

Additional reading: Asimov. *Asimov's new guide to science;* p. 337.

Vanadium CHEMISTRY

A silver-white hard metal, vanadium, atomic number 23, was first discovered in 1801 by Manuel del Rio, a Mexican chemist. Not until 1867 was nearly pure vanadium obtained by the English chemist Sir Henry Roscoe. It occurs in minerals in the western United States and in South Africa. Most of its use is as an additive in the steel industry, giving wear resistance to metals. Chemical symbol is V.

Additional reading: *Van Nostrand's scientific encyclopedia;* p. 2906–2907.

Varnish MATERIALS

Varnish is a solution of a hard resin, drying oil, and a solvent. It is transparent, having no pigments to mask surfaces such as are found in paint. By 1000 B.C. the Egyptians had prepared a varnish using the gum of the acacia tree. Natural resins were used until recent times, when synthetic resins were developed.

Additional reading: *Illustrated science and invention encyclopedia;* v. 19, p. 2544–2545.

Venturi Tube FLUID MECHANICS

When a fluid is moving, the pressure it exerts drops. An explanation for the effect came in 1738 from the Swiss physicist Daniel Bernoulli, who used the law of **conservation of energy.** The Italian physicist Giovanni Venturi developed a device known as the Venturi tube around 1797 which used this effect to measure the rate of flow of a fluid. The principle is used in flow meters, air speed indicators, and related applications.

Additional reading: *Illustrated science and invention encyclopedia;* v. 19, p. 2556–2557.

Venus ASTRONOMY

The ancients were aware of the evening star and the morning star. It wasn't until around 500 B.C. that the Greek philosopher Pythagoras recognized that the morning star and the evening star were one and the same. The first astronomer to study Venus with a **telescope** was the Italian Galileo Galilei in 1610. He observed that it had phases like the moon, showing that it revolved around the sun, as predicted by Copernicus. Later years saw much study of the planet, but its features were not well known until microwaves were reflected off it in 1956 by the team headed by the American astronomer Cornell Mayer. The first of several **space probes** was launched in 1962 and they continued to provide much data.

Additional reading: Asimov. *Asimov's new guide to science;* p. 114–119.

Video Camera ELECTRONICS

The first cameras for video recording were prefessional studio models, based on the equipment invented by Vladimir Zworykin. In 1980 the Sony Corporation (Japan) developed a prototype portable camera containing a built-in videocassette recorder. Many other companies have since produced similar models. (*See also* Television; Videocassette Recorder)

Additional reading: Giscard d'Estaing. *World almanac book of invention;* p. 116.

Video Disk. *See* **Laser Disk.**

Videocassette Recorder
ELECTRONICS

Work on use of magnetic tape for recording images began in the early 1950s. At Ampex Corporation the project was led by Charles P. Ginsberg, Charles E. Anderson, and Ray M. Dolby. The first practical video recorder was manufactured and sold by that company in 1956, for use by television stations, at a cost of about $50,000. In 1958 Ampex built the first color recorder. In 1970 emphasis began to be placed on the use of cassettes for tapes. That year the Philips company announced its first cassette-style model, for which production began in 1972. In 1982 smaller cassettes were made for use in portable systems. The tape for these recorders was first developed at 3M Corporation in 1956 by Melvin Sater and Joseph Mazzitello. (All persons named were U.S.) (*See also* Video Camera)

Additional reading: Giscard d'Estaing. *World almanac book of inventions;* p. 115–116.

Virus
VIROLOGY

In 1892 Russian botanist Dmitri Iosifovich Ivanovsky proved the existence of a pathogenic agent smaller than bacteria while working with tobacco mosaic disease. Ivanovsky happened not to trust the evidence, however. Dutch botanist Martinus Willem Beijerinck observed that an agent other than bacteria was responsible for tobacco mosaic disease, and published his findings in 1898. He called that agent "virus," from the Latin word meaning poison. An American biochemist, Wendell Meredith Stanley, isolated and crystallized a virus for the first time in 1935. The virus in this case was again that of the tobacco mosaic disease. Stanley shared the 1946 Nobel Prize in chemistry for his work with viruses.

Additional reading: Magner. *A history of the life sciences;* p. 280–283.

Virus Culture
VIROLOGY

In 1948 American microbiologists John Franklin Enders, Frederick Chapman Robbins, and Thomas Huckle Weller successfully grew mumps virus on mashed chick embryos. They grew the virus without contamination by bacteria using the recently discovered **penicillin.** The following year they extended their technique and grew polio virus on stillborn human embryo tissue. Until this work, polio virus could be grown only on living human or monkey nerve tissue. This new ability to grow polio virus made it possible to study the virus and contributed to the development of the **polio vaccine.** Enders, Robbins, and Weller shared the 1954 Nobel Prize in medicine and physiology for their work on virus culture.

Additional reading: Sourkes. *Nobel prize winners in medicine and physiology 1901–1965;* p. 315–322.

Viscosity
FLUID MECHANICS

Viscosity refers to the internal frictional forces in a fluid that resist flow. In liquids it is due to intermolecular cohesive forces, while in gases it is due to the movement of molecules in different regions of the gas running at different speeds. Around 1600 the English physicist Sir Isaac Newton formulated more appropriate ideas. Those substances requiring a certain amount of force to be applied before flow occurs are said to have plastic flow. Further developments of Newton's ideas came from the French physician Jean Poiseuille in 1844 and from Sir George Stokes (England) in 1850, the latter showing that objects falling through a viscous liquid reach a steady velocity known as the terminal velocity.

Additional reading: *Illustrated science and invention encyclopedia;* v. 19, p. 2568–2570.

Vitamin A
BIOCHEMISTRY

Elmer Vernon McCollum and Marguerite Davis, American biochemists, found vitamin A in butter and egg yolk in 1913. Thomas Burr Osborne (U.S.) and Lafayette Mendel (U.S.) made the discovery simultaneously, but McCollum was able to publish his and Davis' findings first. In 1915 McCollum labelled the substance "fat-soluble A," a name changed to "vitamin A" in 1920 by British biochemist Jack Cecil Drummond. In 1937 American chemists Harry Nicholls Holmes and Ruth Elizabeth Corbet isolated vitamin A as crystals from cod liver oil.

Additional reading: Asimov. *Asimov's new guide to science;* p. 701–714.

Vitamin B
BIOCHEMISTRY

Edward Vedder (U.S.) and Robert Williams (U.S.) are given credit for being the first to detect water-soluble vitamin B in 1912 as an antineuritic substance effective in curing pigeons of neuritis, a disease similar to that of beriberi in humans. Three years after this discovery, Elmer Vernon McCollum (U.S.) and Marguerite Davis (U.S.) labelled it "water-soluble B," which British biochemist Jack Cecil Drummond changed to "vitamin B" in 1920. Robert Williams (U.S.) synthesized vitamin B-1 in 1934, and came to realize the existence of the B-vitamin complex.

Additional reading: Asimov. *Asimov's new guide to science;* p. 701–714.

Vitamin C
BIOCHEMISTRY

During the 1920s, British biochemist Jack Cecil Drummond suggested that an antiscurvy factor postulated by Polish-born Casimir Funk (U.S.) in 1912 was a substance called vitamin C. A Hungarian-born biochemist, Albert Szent-Gyorgyi (U.S.) succeeded in isolating vitamin C in 1932, and was awarded the 1937 Nobel Prize in medicine and physiology as a result. Simultaneously in 1932, Charles Glen King, an American biochemist, also isolated vitamin C and went on in 1933 to determine its structure. Also in 1933, Tadeus Reichstein (Switzerland) synthesized ascorbic acid. In 1934, Sir Walter Norman Haworth, an English chemist, synthesized it independently of Reichstein and suggested the name ascorbic acid.

Additional reading: Asimov. *Asimov's new guide to science;* p. 701–714.

Vitamin Concept
BIOCHEMISTRY

The Englishman Sir Frederick Gowland Hopkins is given credit for approaching the discovery of the vitamin concept when, in 1906, he determined that food contains essential ingredients beyond carbohydrates, minerals, fats, proteins, and water. Hopkins shared the 1929 Nobel Prize in medicine and physiology for his discovery of growth-stimulating vitamins. Polish-born Casimir Funk (U.S.) made great advances in 1912 when he hypothesized that certain diseases such as beriberi, scurvy, pellagra, and rickets are caused by deficiences of nutrients he called "vitamines."

Additional reading: Asimov. *Asimov's new guide to science;* p. 701–714.

Vitamin D
BIOCHEMISTRY

Elmer Vernon McCollum (U.S.) and his colleagues determined in 1922 that an antiricketic substance existed as a nutrient in cod liver oil. They called this nutrient vitamin D. In 1926 British biochemists Otto Rosenheim and T.A. Webster found that sunlight converted a steroid called ergosterol into vitamin D. At the same time German chemist Adolf Windaus made a similar discovery independently, contributing to his 1928 Nobel Prize in chemistry.

Additional reading: Asimov. *Asimov's new guide to science;* p. 701–714.

Vitamin E
BIOCHEMISTRY

The existence of a fertility substance found in such foods as fresh lettuce, wheat germ, and dried alfalfa leaves was suggested in 1922 by Herbert McLean Evans (U.S.) and K.J. Scott (U.S.). In 1923 Barnett Sure (U.S.) recommended that the substance be labelled vitamin E, and extended the list of foods in which it is found to include polished rice, yellow maize, and rolled oats. Evans and his group were able to isolate vitamin E in 1936, and named it "tocopherol," from the Greek word meaning "to bear children."

Additional reading: Asimov. *Asimov's new guide to science;* p. 701–714.

Vitamin K
BIOCHEMISTRY

Danish biochemist Carl Peter Henrik Dam discovered vitamin K in the 1930s, and found it to be a factor in blood clotting. In 1939 Edward Adelbert Doisy (U.S.) isolated it and determined its structure. Dam and Doisy shared the 1943 Nobel Prize in medicine and physiology for their research.

Additional reading: Asimov. *Asimov's new guide to science;* p. 701–714.

Volcanism. *See* Plutonism.

Volcano
GEOLOGY

Volcanic rocks were not known as such until around 1751, when the French naturalists Chretien Malesherbes and Jean Etienne Gucttard began to study extinct volcanoes in central France. Nicolas Desmarest (France) had realized by around 1764 that lava from volcanoes followed valleys that had been cut by rivers. Little was known then about the source of heat of volcanoes. A very thorough study of French and Italian volcanoes was made in 1827 by the English geologist Sir Charles Lyell. His books published in 1830-1833 showed his mastery of many aspects of geology, including vulcanism, fossils, and stratigraphy. His concepts of the nature of volcanoes contradicted many former conceptions.

Additional reading: Faul. *It began with a stone;* p. 85, 89, 127–134.

Voltmeter
ELECTRICITY

Voltmeters are instruments for measuring voltage and may be one of several types—moving iron, moving coil, electrostatic, and digital. The first two types are similar to **ammeters;** ammeters can be converted to voltmeters by including a large resistance in series. The principle behind them was discovered in 1820 by Hans Christian Oersted (Denmark). The electrostatic type is more recent, using the principle of a mechanical force acting on bodies at different voltages; no current is drawn. The digital style meter was invented in 1952; it uses solid-state circuits. Around 1970 integrated circuits were introduced, for use when high accuracy is required. Multimeters are instruments for measuring voltage, current, and resistance.

Additional reading: *Illustrated science and invention encyclopedia;* v. 19, p. 2578–2580.

Vulcanite
DENTISTRY

Invented in 1851 by Nelson Goodyear (U.S.), this hard rubber substance revolutionized denture construction when Thomas W. Evans (U.S.) began using it in place of gold as a denture base. In 1864 another dentist, John A. Cummings (U.S.), was granted a patent for procedures used in making rubber dentures. In a step that subsequently created great controversy, Cummings sold the patent to the Goodyear Dental Vulcanite Company, which controlled use until 1881, when the patent expired. (*See also* Denture)

Additional reading: Ring. *Dentistry: an illustrated history;* p. 242–243.

Vulcanized Rubber
MATERIALS

As far back as the time of Columbus, South American natives made crude objects out of sap from certain trees. In 1770 the English chemist Joseph Priestly discovered the substance would rub out pencil marks and named it rubber. It was a sticky material until in 1839 when the American inventor Charles Goodyear accidentally dropped a mixture of sulfur and rubber on a hot stove and found that it was soft and pliable even when cold. Adding sulfur to rubber is now called vulcanization, and Goodyear's discovery began the rubber industry, including the manufacture of **rubber tires.**

Additional reading: Asimov. *Asimov's new guide to science;* p. 535.

W

Washing Machine
MECHANICAL ENGINEERING

One of the earliest types of washing machine had a wooden bin for the water in which heavy blades stirred the materials. One was used in an English laundry in 1830. A crank-operated industrial version appeared in France in the 1840s. A similar model was devised in America in 1850 by Joel Houghton. In 1907 an electric washing machine was invented by the American inventor Alva J. Fisher. By the 1930s they were in common use in homes. Around 1950 automatic washers with pre-set programs began to be popular.

Additional reading: Strandh. *A history of the machine;* p. 230–231.

Wasserman Test
IMMUNOLOGY

In 1906 August von Wasserman, a German bacteriologist, developed a diagnostic test for syphilis. The Wasserman test, as it is known, is a complement-fixation test based on serum reaction to certain antigens.

Additional reading: Asimov. *A short history of biology;* p. 127–129.

Waste and Sewage Disposal
CIVIL ENGINEERING

The use of sewers dates back to Assyria (in Nineveh) in 3000 B.C. A large sewer built in Rome around 500 B.C. is still partly used. One method devised in 1895 by Cameron (England) was to put chemically treated sludge in large tanks. In 1916 Ardern (U.S.) and Lockett (U.S.) designed the activated sludge method, which uses biological organisms for purification. Handling of other types of waste, such as paper and garbage, has made use of land fills, pulverization, incineration, and composting. The latter system dates back to the time of the Romans. Incinerators were used as early as 3000 B.C. in Pakistan, as found by Sir R. E. Mortimer Wheeler (England) in the mid-1970s. (*See also* Pollution Control; Toilet)

Additional reading: *Illustrated science and invention encyclopedia;* v. 15, p. 2081–2083, v. 19, p. 2603–2606.

Watch
HOROLOGY

One of the first watches was made in 1504 by the German locksmith Peter Henlein. Watches could not have been invented before the mainspring was created for **clocks** around 1470. Jeweled watches using pivot-holes made of sapphires were first used by the Swiss watchmakers Nicholas Faccio de Duiller and Peter Jacob Debaufe in 1704. The first wristwatch was made in 1790 by Jacquet Droz and Paul Leschot (both Swiss). The first self-winding wristwatch was patented in 1924 by John Harwood (England). In 1930 he invented the first electric watch, but production had to be halted until World War II ended. In 1961 the first electronic watch, using a mercury cell, was invented by the Bulova Company (U.S.) (*See also* Pulse Watch)

Additional reading: *Encyclopedia of inventions;* p. 90–91.

Water
CHEMISTRY

Although humans have used water since the dawn of civilization, its exact chemical composition was not known until 1784, when the English chemist Henry Cavendish discovered that it consisted of hydrogen and oxygen. Later he proved the exact ratio of the two gases.

Additional reading: Downs. *Landmarks in science;* chap. 41.

Water Clock. *See* Clock.

Water Frame
TEXTILES

An invention credited with being one of the foundations of the modern textile industry was the water frame, or the spinning frame, created by the English industrialist Sir Richard Arkwright in 1769. He used water power to operate a spinning machine that was fast and that produced stronger yarn than Hargreaves' **spinning jenny.** Later Arkwright was the first to use a **steam engine** to run a cotton mill (1785).

Additional reading: Feldman. *Scientists & inventors;* p. 62–63.

Water Heater
ENGINEERING

Domestic hot water heaters were originally part of central heating systems, but in time they began to appear as separate units, heated by gas or by electricity. Gas heaters appeared in the 1850s, while electric heaters came later. A solar water heater was invented in France in 1977, and in 1978 David Little (Australia) invented a solar-powered water heater that could rotate its position relative to the sun. Millions of solar heaters are now in use around the world.

Additional reading: *Illustrated science and invention encyclopedia;* v. 9, p. 1191; v. 16, p. 2134–2137.

Water Mill
AGRICULTURE

Two kinds of water mill appeared in the Mediterranean area during the first century B.C. The first was a primitive vertical mill adapted from a large **quern** and powered by water. Although fairly inefficient, it was used by peasants to grind grain. Between 20 and 11 B.C., a Roman engineer, Vitruvius, adapted this water mill and designed the "Vitruvian mill" consisting of a horizontal shaft that carried a vertical water wheel which was geared by cogwheels to a vertical spindle that turned an upper stone. By the fourth century A.D. the Vitruvian water mill was used on a commercial scale to mill grain. (*See also* Water Wheel)

Additional reading: Curwen. *Plough and pasture;* p. 134–142.

Water Pollution. *See* **Pollution Control.**

Water Power. *See* **Hydroelectric Station.**

Water Pump
MECHANICAL ENGINEERING

One of the earliest types of water pumps was invented around 250 B.C. by Archimedes, a Greek mathematician and engineer. It consisted of a metal pipe shaped in the form of a corkscrew, which drew up the water as it revolved. Earlier Egyptian pumps of this sort may have existed. Centuries later a vast improvement was the use of a **steam pump.**

Additional reading: Downs. *Landmarks in science;* chap. 5.

Water Supply. *See* **Aqueduct; Pollution Control; Reservoir.**

Water Wheel
MECHANICAL ENGINEERING

Water power has long been used; it was first evidenced in water wheels that were probably used for turning millstones to grind grain. An early example was located in Turkey around 100 B.C. An important one was discovered in France, built by the Romans around 400 A.D.; it had eight pairs of wheels, used for driving millstones. One of the earliest designs had a vertical shaft for the wheel. Later came the undershot and the overshot designs, depending on which part of the wheel received the flow of water. They used horizontal axles. Some water wheels had gears to change the direction of motion. Another early use was probably that of pumping water. **Steam engines** gradually replaced water mills, beginning in the eighteenth century. (*See also* Hydroelectric Station)

Additional reading: Landels. *Engineering in the ancient world;* p. 17–26.

Wave Mechanics
QUANTUM MECHANICS

The nature of the **electron** was examined by the French physicist Louis de Broglie, and in 1923, using Einstein's formula of mass and energy and Planck's **quantum theory,** he showed that for any particle there should be an associated wave, which he called "matter waves." His theory was verified in 1927 by the American physicists Clinton Davisson and Lester Gerner, who noted that under certain circumstances electron beams would cause diffraction patterns, a phenomenon normally associated with waves. De Broglie's work led the Austrian physicist Erwin Schrodinger to devise in 1926 mathematical expressions to explain such phenomena, a system known as wave mechanics (or quantum mechanics). This work put quantum theory on a firm basis and led to the **electron microscope.**

Additional reading: Feldman. *Scientists & inventors;* p. 282–283.

Weapon. *See* **Cannon; Laser; Machine Gun; Missile; Nuclear Weapon; Revolver; Rifle.**

Weather Forecasting
METEOROLOGY

A knowledge of air masses and their formation and movement was sought by the Norwegian physicist Vilhelm Bjerknes in the 1890s. In the 1930s he proposed that weather conditions were closely related to the formation of warm and cold fronts. He also studied the formation of land and sea breezes as well as the development of high and low pressure areas. His findings led to improved weather forecasting. The use of computers has had a major effect on modern forecasting, as have weather satellites. The latter came into being in 1960 when the United States launched a series of satellites in the Tiros (Television Infrared Observation Satellite) project. These satellites provided photos of cloud coverage, including hurricanes. In 1964 the Nimbus series began, providing night photos also.

Additional reading: Feldman. *Scientists & inventors;* p. 240–241.

Weather Modification
METEOROLOGY

Efforts to modify the weather on a scientific basis date back to the 1940s when the American chemists Vincent Schaeffer and Irving Langmuir studied the role of various nuclei in raindrop formation. One experiment in 1948 involved dropping powdered carbon dioxide into some clouds to serve as nuclei, with rain often resulting from this cloud seeding. Around 1948 the American physicist Bernard Vonnegut found that silver iodide generated on the ground, directed upward, had an even better effect. Other efforts such as seeding hurricanes in order to avert a destructive storm have had only questionable results.

Additional reading: Asimov. *Asimov's biographical encyclopedia;* p. 681–682, 815, 850.

Weather Station
METEOROLOGY

The importance of having a system of regular observations of the weather was evident as far back as 1855 when Napoleon III asked the French astronomer Urbain LeVerrier to set up a network of weather stations, which resulted in ten such stations. By the late 1800s most European countries and the United States had their own systems. Since then thousands of reporting stations have been established all over the world.

Additional reading: Burke. *Connections;* p. 36–38.

Weaving. *See* **Loom.**

Welding
METALLURGY

Several types of welding have been devised over the years. In 1801 Robert Hare (U.S.) developed a flame using **hydrogen** and **oxygen**; other gases such as acetylene have since been used. In 1900 Edmund Fouche (France) developed the first oxy-hydrogen torch. Welding using the heat generated by an electric arc is called arc welding and was devised by Elihu Thompson (U.S.) in 1886. In 1887 carbon rods for arc welding were first developed by M. V. Bernados (Russia). Newer types since devised include electric beam welding (involving emission of **electrons** at a high velocity in a vacuum) and **laser** welding (using a powerful beam of light from a carbon dioxide laser).

Additional reading: *Illustrated science and invention encyclopedia;* v. 20, p. 2649–2654.

Wheel
MECHANICAL ENGINEERING

The exact date of the invention of the wheel will probably never be known, but about 4000 B.C. solid wheels were found to be in use in Sumeria. They were replaced by spoked wheels, which were in fairly common use in Mesopotamia by about 2000 B.C. in Mesopotamia.

Additional reading: Clark. *Works of man;* p. 11–12.

White Blood Cell. *See* **Leukocyte.**

White Dwarf Star
ASTRONOMY

White dwarf stars are very densely packed stars that have passed the stage of being Red Giant stars and are beginning to collapse. As they do so, their core becomes even more dense. Most have a low apparent brightness and are thus difficult to detect in the sky. The first one discovered was Sirius B, which had been predicted by the German astronomer Friedrich Bessel in 1844 on the basis of strange minor periodic motion of Sirius. Not until 1862 was it discovered by the **telescope** builder Alvan Clark (U.S.), although it was not known to be a White Dwarf at that time. In 1914 the American astronomer Walter Adams observed that Sirius B was still hotter than our sun. From its characteristics it was found to be 130,000 times more dense than **platinum**.

Additional reading: Buedeler. *The fascinating universe;* p. 219–222.

Wind Scale. *See* **Beaufort Wind Scale; Petersen Scale.**

Wind Tunnel
AEROSPACE ENGINEERING

The wind tunnel, a necessary device for the aerodynamic testing of a flow of air or other gases around an object, was built in 1871 by the English inventors Francis H. Wenham and John Browning. The Czech-born Ernst Mach also built one in 1887 and gained fame thereby. The tunnels were originally used for the design of **aircraft**, but in time they have also been used for the designs of buildings, railroad cars, **automobiles**, and **missiles**. In more recent years tunnels built for testing supersonic flow resulted in the design of special nozzles for gas entry. Use of the tunnel to determine the conditions at the speed of sound aided engineers in designing a plane capable of **supersonic flight.**

Additional reading: *Illustrated science and invention encyclopedia;* v. 20, p. 2677–2680.

Windmill
MECHANICAL ENGINEERING

The earliest windmills were probably developed in Persia around 700 A.D. and used for milling grain or lifting water. They were put to use in England and France around 1180. Early ones had sails to turn the rotating blades into the wind. Around 1340 windmills in Holland were developed for draining low lands. Tower mills, consisting of a brick base on which sat a movable top portion, were developed in France around 1430 and in England later in the fifteenth century. Small-scale generation of electricity has been in existence for several decades, but no large-scale operation has been successful to date except for a few scattered experimental models.

Additional reading: *Illustrated science and invention encyclopedia;* v. 10, p. 2669–2676; v. 21, section 34.

Wire. *See* **Barbed Wire; Wire Rope.**

Wire Rope
MATERIALS

A wire rope, sometimes called a cable, is made by twisting a number of wires to make a strand, then twisting a number of strands to make the rope. The first such ropes were made in 1832, and in 1840 the first patent for making wire rope was granted to Robert Newell (England).

Additional reading: *Illustrated science and invention encyclopedia;* v. 15, p. 2688–2690.

Wolfram. *See* **Tungsten.**

Word-Association Test
PSYCHOLOGY

Sir Francis Galton, an English anthropologist, was the first person to experiment with association tests in 1879. Galton took words from a dictionary and then recorded the words of which they in turn reminded him. In 1906 Swiss psychiatrist Carl Gustav Jung developed word-association tests designed to reveal the **unconscious** mind by forcing a person to make quick responses, uninhibited by reason. Jung used these tests in his work with psychotic patients.

Additional reading: Murray. *A history of Western psychology;* p. 233, 310–311.

Word Processing
COMPUTER SCIENCE

The process of word processing refers to the use of technically advanced equipment for the storage, editing, and printing of data. It may be accomplished on electronic devices specifically designed for that purpose or on **digital computers** equipped with suitable software and printers. The term is thought to have been originated in the late 1950s by Ulrich Steinhilper (Germany), an employee of IBM. That company used the term in 1967 in marketing its magnetic tape electronic typewriter. By the 1970s the term was expanded to include uses other than typing, such as file updating, scheduling, and list preparation. Word processing is at the heart of "office automation" as a means of making office work speedier and more efficient.

Additional reading: *Illustrated science and invention encyclopedia;* v. 20, p. 2701–2704.

Writing Material
GRAPHIC ARTS

Aside from clay tablets, used by the Mesopotamians over 5,000 years ago, the papyrus reed found along the Nile River became the best writing material when discovered for this use around 2500 B.C. Criss-crossed layers were beaten, then dried in the sun. Around 200 B.C. skins of new-born lambs or calves were found to make a fine writing material called parchment. By around 100 A.D. parchment was used in rolls rather than sheets. The manufacture of paper began in China around 100 A.D., using compacted plant fibers. Mohammedan conquerors in the eighth century spread papermaking throughout Islamic countries and then to Europe by around 1150. Linen waste was used to produce rag paper by 1300. The first papermaking machine was patented in 1799 by Nicolas Robert (France).

Additional reading: *Illustrated science and invention encyclopedia;* v. 21. section 18.

X

X Chromosome. *See* **Chromosome.**

X-ray Astronomy
ASTRONOMY

The fact that **neutron stars** emit **X-rays** along with other radiation has led to the development of X-ray astronomy. It had its beginning in the United States in the early 1960s when a search began using high-altitude **rockets** and **satellites** for the source of X-rays previously observed. Two of the astronomers in the early years were the Italian-born Bruno Benedetto (U.S.) and Herbert Friedman (U.S.), who discovered the first X-ray stars in 1963. In ten years more than 100 such stars were found. Stars emitting gamma rays led to gamma-ray astronomy; **radio astronomy** began as early as the 1930s.

Additional reading: Buedeler. *The fascinating universe;* p. 226–227.

X-ray Crystallography
CRYSTALLOGRAPHY

Proof that **X-rays** were waves, not particles, was given in 1912 when the German physicist Max von Laue exposed some crystals to a beam of X-rays. The resulting pattern of bright dots was caused by the layers of atoms in the crystal, producing the familiar diffraction patterns formed when using light waves. His work led to the use of X-rays to study the structure of unknown crystals, primarily carried on around 1912 by Sir William Lawrence Bragg (England), and his father, William Henry Bragg (England). In 1915 the Braggs shared the Nobel Prize in physics, while Laue won it in 1914.

Additional reading: Asimov. *Asimov's new guide to science;* p. 272–273.

X-rays
NUCLEAR PHYSICS

Through accident the existence of X-rays became known to the German physicist Wilhelm Roentgen in 1895. He gave them this name because he didn't know what they were, X standing for the unknown. The discovery of **radioactivity** was said to be a result of the discovery of X-rays. They are widely used in medicine and in industrial examinations of materials.

Additional reading: Asimov. *Asimov'e biographical encyclopedia;* p. 502–504.

X-rays (Medicine)
MEDICINE

In 1895, soon after the discovery of **X-rays,** physicians began to examine bone fractures with them. Walter Bradford Cannon, an American physiologist, was the first person to apply X-rays for physiological purposes. In 1896 he fed bismuth subnitrate to a small animal and followed its course through the digestive system with the aid of a fluorescent screen and X-rays.

Additional reading: Woglom. *Discoverers for medicine;* p. 160–176.

Xenon
CHEMISTRY

Xenon, atomic number 54, is normally a colorless, odorless gas. It was discovered in 1898 by the Scottish chemist Sir William Ramsay and the English chemist Morris Travers, by fractional distillation of krypton. It forms stable compounds only under certain conditions, such as excitation in discharge tubes or under pressure. It is present in a very small quantity in air, and is produced as a by-product in the liquefaction and fractional distillation of air. Its chief use is as a filling gas for light bulbs and photographic flash lamps as well as in bubble chambers and lasers. Chemical symbol is Xe.

Additional reading: *Van Nostrand's scientific encyclopedia;* p. 3039–3040.

Xerography
GRAPHIC ARTS

A desire for a quick, clean, and inexpensive method of making copies of printed material led the American technician Chester Carlson to develop in 1938 an electrostatic method, which became known as xerography (derived from the Greek meaning "dry writing"). The basic idea involves a selenium drum on which an image of the page is reflected, the formation of an electrically charged copy of the image in powder form, then the fusing of the powder on the paper by application of heat. It was not until 1947 that Carlson got financial backing, leading to the eventual outstanding success of the Xerox Corporation. Since then color copiers as well as enlargement/reduction features have been developed.

Additional reading: Feldman. *Scientists & inventors;* p. 298–299.

Y

Y Chromosome. *See* **Chromosome.**

Yeast
MYCOLOGY

In 1680 the Dutch biologist and microscopist Anton van Leeuwenhoek observed yeast cells under a single lens **microscope** and described their structure. Yeast was found to be a living organism in 1836 by Charles Cagniard de la Tour (France) and Theodor Schwann (Germany).

Additional reading: Asimov. *A short history of biology;* p. 90–92.

Young's Modulus. *See* **Modulus of Elasticity.**

Ytterbium
CHEMISTRY

A silver-gray color in metallic form, ytterbium, atomic number 70, is one of the least abundant elements of the rare-earth group. It was discovered in 1878 by the Swiss chemist Jean-Charles de Marignac while studying some rare-earth metals. The chief sources are minerals such as euxenite, gadolinite, monazite, and xenotime, of which the latter is most important. It has been used in infrared devices, in lasers, and in radiographic units. Chemical symbol is Yb.

Additional reading: *Van Nostrand's scientific encyclopedia;* p. 3054–3055.

Yttrium
CHEMISTRY

Yttrium, atomic number 39, is a silver-gray color in metallic form. It was discovered in 1794 by the Finnish chemist Johan Gadolin. It is normally obtained by recovery from rare-earth minerals and also as a by-product of **uranium** mining operations in Canada. Xenotime in Malaysia is also a good source. The main use has been in **nuclear reactor** hardware and more recently in **color television** picture tubes. Chemical symbol is Y.

Additional reading: *Van Nostrand's scientific encyclopedia;* p. 3055.

Z

Zeppelin. *See* **Airship.**

Zinc
CHEMISTRY

Zinc, atomic number 30, has been known since around 500 B.C., as zinc-containing ornaments of that era have been found. Early smelting and purification processes took place in China around 1000 A.D. It is found in conjunction with minerals containing **lead** and **copper**, with major deposits widely spread around the world. One of its chief uses is in galvanizing to protect metals against corrosion, particularly **iron** and **steel.** Another is in castings for commercial manufactured products. Chemical symbol is Zn.

Additional reading: *Van Nostrand's scientific encyclopedia;* p. 3059–3062.

Zipper
MECHANICAL ENGINERING

The first attempt at making a zipper came in 1893 when Whitcomb Judson (U.S.) patented a closure having a movable slider. In 1905 he improved on the first model, but it was not until 1913 that Gideon Sundback (Sweden) devised a zipper having interchangeable and identical teeth, used in 1918 for aviation suits. In 1920 the B. F. Goodrich Company put them in galoshes.

Additional reading: *Illustrated science and invention encyclopedia;* v. 20, p. 2729–2730.

Zirconium
CHEMISTRY

Crystalline zirconium, atomic number 40, is a white, soft, ductile, and malleable metal. It was discovered in 1789 by the German chemist Martin Klaproth. It is found chiefly in zircon, often located on beaches and lake beds. It is normally obtained as a by-product of the processing of ilmenite and rutile. One of its main uses is in water-cooled **nuclear reactors** (because of its high corrosion resistance) and also in chemical processing equpment. Chemical symbol is Zr.

Additional reading: *Van Nostrand's scientific encyclopedia;* p. 3064–3065.

Zone Refining. *See* **Crystal Growth.**

Zoology
BIOLOGY

The Greek philosopher Aristotle, who died in 322 B.C., is credited with founding the science of zoology. Aristotle was particularly interested in natural history, and developed a classification system for animals. In addition, he made observations on marine biology and embryology. Aristotle established the Lyceum in Athens, at which an emphasis was placed on the study of biology and history. (*See also* Animal and Plant Classification)

Additional reading: Sarton. *A history of science;* p. 529–545.

Listing of Cited References

The following reference sources constitute a listing of all the books (and a few periodical articles) that were referred to in the entries in the "Additional reading" section.

Ackerknecht, Erwin H. *A short history of medicine.* Rev. printing. New York: The Ronald Press Company; 1968. 275p.

A concise history of medicine, this book is aimed toward both medical professionals and general readers interested in core areas of the discipline. The book begins with primitive medicine and moves into the twentieth century. A special chapter devoted to the history of medicine in the United States is included. A good list of suggested readings is provided.

Alexander, Franz G.; Selesnick, Sheldon T. *The history of psychiatry: an evaluation of psychiatric thought and practice from prehistoric times to the present.* New York: Harper & Row; 1966. 471p.

The authors, accomplished psychoanalysts, have written a scholarly history of psychiatry that begins in Mesopotamia and covers the Ancients and subsequent periods up to the mid-twentieth century. A large portion of the discussion focuses on the Freudian Age. A lengthy bibliography is included.

Arnow, L. Earle. *Health in a bottle: searching for the drugs that help.* Philadelphia: J. B. Lippincott; 1970. 272p.

Dr. Arnow, former director of research at three pharmaceutical companies, is well qualified to give an insider's view of the processes involved in drug research and production. He describes the ways in which new drugs are discovered and tested, the role played by animal and human "guinea pigs," and the steps necessary for compliance with FDA, patent, and trademark regulations.

Asimov, Isaac. *A short history of biology.* Garden City, NY: The Natural History Press; 1964. 182p.

This is a concise, authoritative account of the development of biology from the ancient Greeks to modern day. Asimov writes in an entertaining style that helps general readers over the more difficult developments of molecular biology, genetics, and the like.

———. *Asimov's biographical encyclopedia of science and technology: the lives and achievements of 1510 great scientists from ancient times to the present chronologically arranged.* 2d rev. ed. New York: Doubleday; 1982. 941 p.

A remarkable book, giving detailed accounts of the scientists (and many engineers) named. The work has been carefully and accurately done. There is an index by personal names, as well as by subject. An important biographical tool.

———. *Asimov's new guide to science.* New York: Basic Books; 1984. 940 p.

An excellent review of developments in all branches of science, ranging from ancient times to the present. Gives names of scientists, dates, and locales for all major events. Has excellent indexes. An indispensable reference work.

———. *The history of physics.* New York: Walker & Company; 1984. 720 p.

A thorough discussion of the development of physics, done in a thorough, careful style. Well-documented facts are plentiful in this useful reference source.

Bagrow, Leo. *History of cartography.* Rev. by R. A. Skelton. Cambridge, MA: Harvard University Press; 1960. 312 p.

A thorough history of the making of maps, ranging from ancient times to the eighteenth century. Contains more than 200 maps, some in color.

Baker, David. *Conquest: a history of space achievements from science fiction to the shuttle.* London: Holland & Clark; 1984. 191p.

A well-written history of the world's activities in exploring space. It is beautifully illustrated, almost all in color. Has ten useful appendixes or tables, including a chronological list of manned space slights, lists of astronauts, list of unmanned flights, etc. An excellent book.

Bennion, Elisabeth. *Antique medical instruments.* London: Sotheby Parke Bernet; Berkeley: University of California Press; 1979. 355p.

This beautifully illustrated volume focuses primarily on English surgical instruments developed prior to 1870. Where possible, Bennion includes information on the inventors, along with data on function and construction. She also keeps the reader informed of the historical context in which significant advances occurred. A directory of surgical instrument makers provides names, addresses, and dates in the time period covered.

Berger, Kenneth W. *The hearing aid: its operation and development.* Livonia, MI: The National Hearing Aid Society; 1984. 299p.

Berger has compiled a comprehensive source of information on the electronic design and acoustic responses of hearing aids. It also presents historical and technical hearing aid model data. Treatment begins with pre-electric and nonelectric hearing aids, and includes a discussion of legislation and codes. A hearing aid industry directory is also provided. Photographs and illustrations supplement the text.

Bettmann, Otto L. *A pictorial history of medicine.* Springfield, IL: Charles C. Thomas; 1956. 318p.
This is a nontechnical survey of medicine from the practitioners of Aesculapius to Paul Ehrlich and the advent of chemotherapy in the nineteenth century. As the title suggests, the emphasis is on illustrating this succinct history of the events and concepts important to medicine. Arrangement is chronological.

Bick, Edgar M. *Source book of orthopaedics.* 2nd ed. Baltimore: The Williams & Wilkins Co.; 1948. 540p.
This comprehensive history of orthopaedics begins with primitive man and traces the development of orthopedic surgery to the 1940s. Chapters cover such areas as pathology, bone surgery, and fractures. A bibliography appears at the end of each chapter.

Bracegirdle, Brian. *A history of microtechnique.* Ithaca, NY: Cornell University Press; 1978. 359p.
Bracegirdle's history covers the evolution of the microtome and the development of histological methods. Generously illustrated with photographs and drawings, it also includes a chronological summary listing first uses of synthetic dyes, along with the year and name of the scientist who first used it.

Buedeler, Werner. *The fascinating universe: the modern aspects of astronomy.* Translated by Fred Bradley. New York: Van Nostrand Reinhold; 1983. 252 p.
A thorough discussion of all aspects of astronomy. Includes historical developments. Has many photographs, some in color.

Burke, James. *Connections.* Boston: Little, Brown; 1979. 304 p.
Discusses the changes brought about by various inventions since early times up to the present period, written in an interesting style and profusely illustrated (often in color).

Butterfield, Herbert. *The origins of modern science 1300-1800.* Rev. ed. New York: The Free Press; 1965. 255p.
This compilation of lectures presented at Cambridge in the 1940s addresses the origins of modern science in the context of the history of western civilization. The interplay between beliefs of the day and scientific progress is well explicated by the author. A list of suggested readings is provided.

Bynum, W. F. and others, ed. *Dictionary of the history of science.* Princeton, NJ.: Princeton University Press; 1981. 494 p.
Consists of 700 articles covering major topics in science. Names and dates for inventors are usually included. There is a name index (including dates, nationality, and areas of science).

Carson, Rachel. *Silent spring.* Boston: Houghton Mifflin Co.; 1962. 368p.
Biologist Carson began collecting data in 1958 on the deadly effects synthetic pesticides were having on the environment. She presents her findings in this book and arms general readers with sufficient information to assess for themselves the impact pesticides have on the plant and animal world.

The chemistry of life: eight lectures on the history of biochemistry. Edited by Joseph Needham. Cambridge, England: University Press; 1970. 213p.
Lectures given by Cambridge scientists from 1958 to 1961, covering events occurring from roughly 1800 onward, are herein compiled. Contributors are R. Hill, M. Dixon, E.F. Gale, K. Dixon, F.G. Young, L.J. Harris, M. Teich and R. Peters.

Clark, Ronald W. *Works of man.* New York: Viking; 1985. 351 p.
A readable review of the nature and significance of technical developments from ancient times to the present, written for the layperson. Has many illustrations (most in color).

Cowles encyclopedia of science, industry and technology. New enlarged edition. New York: Cowles Book Company; 1969. 639 p.
A large, detailed account of the history and development of scientific discoveries and inventions. Arranged by broad categories such as life science, space sciences, energy, and materials. Has an index of over 10,000 entries.

Crapo, Lawrence. *Hormones: the messengers of life.* New York: W.H. Freeman & Co.; 1985. 194p.
This is an excellent introduction to the topic of hormones. The information is presented clearly and well supplemented by drawings. The bibliography is also useful.

Curwen, E. Cecil; Hatt, Gudmund. *Plough and pasture: the early history of farming.* New York: Henry Schuman; 1953. 329p.
A particularly helpful source for information regarding the origins of agriculture and animal husbandry. Divided into two sections, the book covers prehistoric farming in Europe and the Near East and the economic culture of non-Europeans as it relates to agriculture. A useful bibliography is provided with each section.

Dale, Tom; Carter, Vernon Gill. *Topsoil and civilization.* Norman: University of Oklahoma Press; 1955. 270p.
The authors analyze the relationship between civilization and the way people exploit the soil for food production. The book has an interesting historical approach that covers Europe, the Middle East, and the United States.

Daumas, Maurice, ed. *A history of technology & invention; progress through the ages.* Translated by Eileen B. Hennessy. New York: Crown; 1962–1969. 2 vols.
Volume one of this set begins with primitive societies and extends to the fourteenth century. Volume two covers the period from the fifteenth through the first part of the eighteenth century. Subjects are treated in a very thorough fashion with a number of drawings and photographs.

Davis, Audrey B. *Medicine and its technology: an introduction to the history of medical instrumentation.* Westport, CT: Greenwood Press; 1981. 285p. (Contributions in Medical History; no. 7.)
This is a nicely illustrated history of medical technology arising in the nineteenth and twentieth centuries. Organization is topical and includes discussions of events such as the development of stethoscopes.

Davis, Joel. *Endorphins: new waves in brain chemistry.* New York: Dial Press; 1984. 275p.
This book is a good introduction to the discovery, research, and possible future uses of endorphins, substances frequently referred to as "brain opiates." The author describes where they are found in the body and brain and how they function. Although the book lacks a subject index, it does include a chronology of important events in brain and endorphin research, along with a lengthy list of sources.

De Bono, Edward, ed. *Eureka!: an illustrated history of inventions from the wheel to the computer.* New York: Holt, Rinehart & Winston; 1974. 248 p.

Offers description of hundreds of inventions, ranging from the sundial to rockets. Arranged in fourteen chapters devoted to such topics as tools, clothing, communication, etc. Profusely illustrated, some in color.

De Camp, L. Sprague. *The ancient engineers.* New York: Doubleday; 1963. 408 p.

A very detailed study of the accomplishments of engineers, ranging from the crude inventions of the Stone Age to around the seventeenth century. Arranged by regions or countries, such as Egypt, Greece, Orient, and Europe. An excellent source, with a fine index.

Dean, R.T. *Lysosomes.* London: Edward Arnold; 1977. 52p. (Institute of Biology's Studies in Biology; no. 84)

The author intends this book to be an introduction to the structure and functions of lysosomes, and their role in pathology. The history of their discovery is briefly addressed. A list of further reading is included.

Derry, T. K.; Williams, Trevor I. *A short history of technology: from the earliest times to A. D. 1900.* New York: Oxford University Press; 1970. 783 p.

A very detailed account, divided into chapters devoted to such topics as food production, working of metals, transportation, materials, etc. Has numerous black and white illustrations. A timescale table covers from 3500 B.C. to 1900. A thorough treatment.

Dictionary of scientific biography. New York: Scribner's; 1970-1980. 16 vols.

A very thorough set of biographies of deceased scientists. Technology and applied science are not included unless the persons also worked in the sciences. There is a separate volume containing person and subject indexes, both very well done. A most useful source of data.

Downs, Robert B. *Landmarks in science: Hippocrates to Carson.* Littleton, CO: Libraries Unlimited; 1982. 305 p.

Consists of 74 chapters, each devoted to the accomplishments of scientists in a wide variety of disciplines. Written for the layperson.

Dummer, G. W. A. *Electronic inventions 1745–1976.* New York: Pergamon; 1977. 158 p.

Describes hundreds of inventions involved with electronics, including computers, components, sound reproduction, etc. Data are listed chronologically, then by name of the invention, with source of the information cited, such as books, periodical articles, etc. There are indexes by inventors' names, by name of the invention, and by broad category (television, computers, etc.).

Dunn, L.C. *A short history of genetics: the development of some of the main lines of thought: 1864–1939.* New York: McGraw-Hill; 1965. 261p.

Dunn covers genetics from its inception with Mendel through to the end of its classical period in 1939. The book treats the beginnings of population genetics and the development of the gene theory. A glossary and long bibliography are included.

———. *Genetics in the 20th century: essays on the progress of genetics during its first 50 years.* Edited for The Genetics Society of America. New York: Macmillan Co.; 1951. 634p.

This is a compilation of papers presented at the Golden Jubilee of Genetics at Ohio State University, Columbus, September 1950, in honor of the 50th anniversary of the rediscovery of Mendel's work. The papers review the development of Mendelian genetics, gene theory, and heredity.

Encyclopaedia Britannica. 15th ed. Chicago: Encyclopaedia Britannica, Inc.; 1982. 30 vols.

A well-known reference source that is useful for providing sci-tech data on many topics of interest to the generalist.

Encyclopedia of inventions. Secaucus, NJ: Chartwell Books; 1976. 128 p.

A useful record of thousands of inventions, providing a detailed text along with hundreds of photographs and drawings. Usually identifies dates and names of inventors. Covers both domestic and commercial products. Has a detailed index.

Engineers and inventors. Edited by David Abbott. New York: Peter Bedrick Books; 1986. 188 p.

A useful compilation of short biographical sketches of some 200 important engineers and inventors, arranged alphabetically by name. There is an index arranged by name as well as by invention.

Fairbridge, Rhodes W., ed. *Encyclopedia of oceanography.* New York: Reinhold; 1966. 1021p. (Encyclopedia of Earth Sciences Series, Vol. 1)

A thorough treatment of all aspects of the subject, including related topics of such disciplines as meteorology, chemistry, etc. It has many charts and drawings.

Faul, Henry; Faul, Carol. *It began with a stone: a history of geology from the Stone Age to plate tectonics.* New York: Wiley; 1983. 230 p.

An interesting history of geology which is quite thorough in its treatment, yet has a style that appeals to the average reader. Dates and names of scientists are carefully documented.

Feldman, Anthony; Ford, Peter. *Scientists & inventors.* New York: Facts on File; 1979. 336p.

Describes the contributions made by more than 150 inventors and scientists from Aristotle to James Watson. Copiously illustrated (many in color), it is aimed at the layperson.

Freiberger, Paul; Swaine, Michael. *Fire in the valley: the making of the personal computer.* Berkeley, CA: Osborne/McGraw-Hill; 1984. 288p.

An engaging account of the people and events involved in the rapid growth of the microcomputer industry in the 1970s and 1980s, including both hardware and software. Very readable.

Friedman, Richard S. and others. *Advanced technology warfare: a detailed study of the latest weapons and techniques for warfare today and the 21st century.* New York: Harmony Books; 1985. 208 p.

Having lavish illustrations (many in color), this reference book describes modern weapons and warfare techniques. Covers all phases, such as electronic warfare, air warfare, space warfare, and land warfare. Very current data.

Garard, Ira D. *Invitation to chemistry.* New York: Doubleday; 1969. 420 p.

Written for the layperson with little or no background in chemistry. Has chapters on such topics as alchemy, early years of chemistry, the elements, carbon, and analytical chemistry.

Giscard d'Estaing, Valerie-Anne. *World almanac book of inventions.* New York: World Almanac; 1985. 361 p.

Aimed at a wide audience, this book contains around 2,000 references to technical and scientific inventions.

Most entries are brief. There is a partial index. Has many illustrations.

Gorin, George. *History of ophthalmology.* Wilmington, DE: Publish or Perish; 1982. 630p.

This comprehensive work reviews the history of ophthalmology from antiquity to 1975. It contains short biographies of individuals and discusses their work and lives. Nearly 300 references are listed.

Greene, Edward Lee. *Landmarks of botanical history.* Edited by Frank N. Egerton. CA: Stanford University Press; 1983. 2 vols.

This is a scholarly and comprehensive history of botany before 1700. Chapters look in depth at individuals and their contributions to the field.

Gribbin, John. *In search of the double helix: quantum physics and life.* New York: McGraw-Hill; 1985. 369p.

Gribbin has written an enjoyable history covering the period from the first discovery of nucleic acids to the establishment of the structure of DNA. This is an excellent source for introductory reading. Gribbin provides an annotated bibliography for the topic.

———. *Our changing planet.* New York: Crowell; 1977. 165p.

Discusses the old and the new ideas about the earth—its formation, its structure, its environment, etc. Written in an interesting style, fully understandable to the layperson. Emphasizes some concepts such as continental drift and plate tectonics.

Hawks, Ellison; Boulger, G. S. *Pioneers of plant study.* New York: Macmillan; 1928. 288p.

This is an historical and biographical account of persons important to the field of botany from days of antiquity through to the mid-1800s. One unique feature is the provision of portrait illustrations for many of the individuals. The material is written with the general reader in mind.

Howells, John G. *World history of psychiatry.* New York: Brunner/Mazel; 1975. 770p.

This a compilation of 29 chapters written by contributors who trace the development of psychiatry from country to country. Included are such areas as Italy, Hungary, Poland, Turkey, Canada, South Africa, India, and the Far East.

Hughes, Arthur. *A history of cytology.* London; New York: Abelard-Schuman; 1959. 158p.

The development of microscopical observation begins this authoritative discussion. The text covers recognition of the cell and theories regarding its function, cell division, formation of germ cells, theories of inheritance, and a history of the study of cytoplasm.

Hurlbut, Cornelius S., Jr. *The planet we live on: illustrated encyclopedia of the earth sciences.* New York: Harry Abrams; 1976. 527p.

A large reference work written for the layperson, having clearly written entries with many drawings and photographs (some in color). A quick source for information on any of the earth sciences.

Hyams, Edward. *Soil and civilization.* New York: Thames and Hudson; 1952. 312p.

This study on the history of man's relationship with the soil covers such topics as the rise of Athens, Hannibalic, the ruin of soil in Oklahoma, the Inca Empire.

Illustrated science and invention encyclopedia: how it works. Westport, CT: Stuttman; 1983. 24 vols.

A copiously illustrated encyclopedia (mostly in color), written for the layperson. A well-prepared, useful reference source with a good index. Unusually strong in the coverage of technology.

Judson, Horace Freeland. *The eighth day of creation: makers of the revolution in biology.* New York: Simon and Schuster; 1979. 686p.

This is a history of the discoveries of molecular biology, how they were made, and the scientists who were responsible. It includes work done in the 1930s and moves up through RNA research done in the 1970s. Jargon has been avoided so that general readers can easily comprehend the material.

Kalckar, Herman M. "The discovery of hexokinase," *Trends in Biochemical Sciences.* 10(7): 291-293; 1985.

The discovery of hexokinase is reviewed. The discussion includes data on the research leading up to its discovery and describes the discovery itself. The language is technical.

Landels, J. G. *Engineering in the ancient world.* Berkeley & Los Angeles: University of California Press; 1978. 224p.

Aims at discussing the technological achievements of the Greeks and Romans. Chapters are devoted to such topics as energy sources, water pumps, cranes, ships, and land transportation. One chapter reviews the writings of prominent Greek and Roman writers.

Lewin, Roger. "The birth of recombinant RNA technology." *Science.* 222: 1313–1315; 1983.

Lewin's article tells about the revolutionary development of a technique allowing large scale production of RNA for the first time. It includes a diagram of the first recombinant RNA molecule.

MacFarlane, Gwyn. *Alexander Fleming: the man and the myth.* Cambridge, MA: Harvard University Press; 1985. 304p.

This biography of Alexander Fleming provides information about the discovery of penicillin in nontechnical language. Interesting facts about Fleming add to the book's appeal. It includes a bibliography of his important papers, plus a lengthy list of suggested readings.

Magner, Lois N. *A history of the life sciences.* New York: Marcel Dekker; 1979. 489p.

Magner's volume offers excellent discussions on the development of biology and focuses on themes important to the field. Individual chapters are included on cell theory, physiology, evolution, and genetics, among others.

Marx, Jean L. "The 1986 Nobel Prize for physiology or medicine." *Science.* 234: 543–544; 1986.

The 1986 Nobel Prize was awarded to Rita Levi-Montalcine and Stanley Cohen for discoveries of factors that control cell growth and development. A concise description of their work and its significance is provided in this report.

McCarty, Maclyn. *The transforming principle: discovering that genes are made of DNA.* New York: W.W. Norton & Co.; 1985. 252p.

This is McCarty's version of the discovery that DNA carries genetic information he made with Oswald T. Avery and Colin M. MacLeod. The book provides a fascinating look at a crucial era in genetics.

McCorduck, Pamela. *Machines who think; a personal inquiry into the history and prospects of artificial intelligence.* San Francisco: Freeman; 1979. 375 p.

Presents a fascinating history of artificial intelligence, written largely in terms of the important people involved in its development. Contains a wealth of information.

McGraw-Hill encyclopedia of science and technology. 5th ed. New York: McGraw-Hill; 1982. 15 vols.

A well-written encyclopedia covering all fields of science and engineering. Has 7,700 entries plus over 15,000 illustrations. There is a separate index volume. It is aimed at the nonspecialist.

McGrew, Roderick E. *Encyclopedia of medical history.* New York: McGraw-Hill Book Co.; 1985. 400p.

This is an excellent starting point for gathering historical information on important medical topics. Essays range from one page to several, and treat the subject matter chronologically. While biographies are not included, individuals' contributions are discussed. Suggested readings are provided.

McIntosh, Robert P. *The background of ecology: concept and theory.* Cambridge; New York: Cambridge University Press; 1985. 383p.

McIntosh accounts for the background of ecology, relating its development to its current status as a science. While not purely historical, the work does provide a very good discussion of individuals and ideas that helped shape the field. A lengthy list of references is included.

McKinnell, Robert Gilmore. *Cloning of frogs, mice, and other animals. Rev. ed. of "Cloning: a biologist reports."* Minneapolis: University of Minnesota Press; 1985. 127p.

This book is appropriate as an introduction to and a concise account of recent advances in cloning technology. A glossary is included, along with a helpful list of references.

Medawar, P.B.; Medawar, J.S. *Aristotle to zoos: a philosophical dictionary of biology.* Cambridge, MA: Harvard University Press; 1983. 305p.

This volume is a compilation of concise articles that discuss significant topics in the biological sciences. Topics include adaptation, bioengineering, creationism, meiosis, toxins and zoos. It provides thumbnail sketches that can assist in defining a given issue.

Morton, A.G. *History of botanical science: an account of the development of botany from ancient times to the present day.* London; New York: Academic Press; 1981. 474p.

The central theme of this work is the evolution of botanical theory from the time of the Romans to the 20th century. It is aimed toward both general readers and botanists and so contains minimal technical language.

Murray, David J. *A history of western psychology.* Englewood Cliffs, NJ: Prentice-Hall; 1983. 428p.

Intended as a textbook, this work provides an excellent introduction to the roots of present-day psychological theory. Coverage begins with the ancients and moves through medieval psychology on up to 1980. Chapters are devoted to such areas as gestalt psychology, behaviorism, and psychoanalysis. An extensive list of references is included.

Newman, James R., ed. *The world of mathematics.* Vol. 1. New York: Simon and Schuster; 1956. 724 p.

Part of a monumental four-volume set, this volume contains over thirty essays devoted to the history and biography of mathematics. Very carefully done, this contains scholarly, factual papers.

Nossal, G.J.V. *Reshaping life: key issues in genetic engineering.* New York: Cambridge University Press; 1985. 158p.

This book presents basic issues of genetic engineering in a style designed for easy comprehension by general readers. Both science and social issues are examined. A glossary and suggestions for further reading are included.

Olby, Robert. *The path to the double helix.* Seattle: University of Washington Press; 1974. 510p.

This comprehenisve history of molecular biology begins with the first discovery of nucleic acids and proceeds up to the establishment of the structure of DNA. Over twenty black and white photographs are included, along with a useful list of references.

Portugal, Franklin H.; Cohen, Jack S. *A century of DNA: a history of the discovery of the structure and function of the genetic substance.* Cambridge, MA: MIT Press; 1977. 384p.

This useful source of information for discoveries about DNA takes coverage up to the solving of the genetic code. The text is supplemented with photographs and charts.

Reiser, Stanley Joel. *Medicine and the reign of technology.* New York: Cambridge University Press; 1978. 317p.

The history of medical instrumentation and technique is presented in an interesting and informative fashion. The book begins with the seventeenth century and moves up to the age of telecommunications and automation. The text is complemented by apt illustrations and a lengthy bibliography.

Ring, Melvin E. *Dentistry: an illustrated history.* New York: Harry N. Abrams; 1985. 319p.

This beautifully illustrated history traces the development of dentistry from the primitive world to the twentieth century. It is an excellent source of information on instrumentation and technique.

Ritchie, David. *The computer pioneers: the making of the modern computer.* New York: Simon and Schuster; 1986. 238 p.

An excellent history of the people and events involved in the development of computers. It begins with the crude predecesors of computers and takes us up to the 1970s. An appendix offers a compilation of thumbnail sketches of major computers; another feature is a brief glossary.

Robinson, Donald. *The miracle finders: the stories behind the most important breakthroughs of modern medicine.* New York: David McKay Co.; 1976. 332p.

The author spent five years interviewing individuals central to the development of modern medicine. His essays give the flavor of medical research and tell some of the personal stories behind the scene. A lengthy bibliography is included.

Rothfeder, Jeffrey. *Minds over matter.* New York: Simon & Schuster; 1985. 238p.

A current book that covers all aspects of artificial intelligence, including expert systems and related topics. Written to be understandable to the layperson.

Sarton, George. *A history of science: ancient science through the golden age of Greece.* Cambridge, MA: Harvard University Press; 1952. 646p.

This is an authoritative history of ancient science in the western world. It treats mathematics, astronomy, physics, and biology. Sarton's works are mandatory reading for anyone interested in the history of science.

Schlebecker, John T. *Whereby we thrive: a history of American farming, 1607–1972.* Ames: Iowa State University Press; 1975. 342p.

Conveniently organized for quick reference, this history covers social, economic, and technological aspects of American farming. Sections are arranged chronologically, with topics organized into discussions about the land, markets, technology, and science. An extensive bibliography concludes the book.

Scott, Joan E. *Introduction to interactive computer graphics.* New York: Wiley; 1982. 255 p.

The hardware section has four chapters on input devices, display screens and secondary devices, while the software section discusses techniques of building programs. Two chapters deal with applications.

Sikora, Karol; Smedley, Howard M. *Monoclonal antibodies.* Boston: Blackwell Scientific Publications; 1984. 132p.

The story behind monoclonal antibodies and their applications in biology and medicine is told in a clear, concise way. A short glossary and well-planned photographs and drawings will aid readers who are not specialists.

Sneader, Walter. *Drug discovery: the evolution of modern medicines.* New York: Wiley; 1985. 435p.

A strength of this volume is that it brings together in one place facts behind the development of major drugs that have significantly affected modern society. While some of the material is technical and includes the provision of chemical formulae, it is accessible to the general reader. An extensive bibliography is included.

Sourkes, Theodore L. *Nobel Prize winners in medicine and physiology 1901–1965.* Rev. of earlier work by Lloyd G. Stevenson. New York: Abelard-Schuman; 1966. 464p.

For each year in which a Prize was awarded, biographical entries, a description of the Prize-winning work, and a discussion of its significance is given. An appendix lists the laureates by the subject of their research.

Strandh, Sigvard. *A history of the machine.* New York: A & W Publishers; 1979. 240 p.

A large-sized, attractively illustrated book which deals exclusively with machines of all types, ranging from wind mills to robots, from the steam engine to computers. Many machines are illustrated, usually with very clear drawings (many in color). The text is clear and includes many details. Highly recommended.

Sturtevant, A.H. *A history of genetics.* New York: Harper & Row; 1965. 165p.

The development of genetics is traced from Mendel to about 1950. A concise chronology is included in an appendix. There is also a section of diagrams that show teacher-student relationships among individuals involved in the study of genetics.

"20 discoveries that changed our lives." *Science. 84* 5(9): 4–190; 1984 November.

An exceptionally well-illustrated presentation of twenty discoveries in science, technology, and medicine considered to be among the most significant of this century. The articles were written by leading scientists and historians, some of whom were involved in the events described. The material is presented in a concise and interesting style.

Valenstein, Elliot S. *Great and desperate cures: the rise and decline of psychosurgery and other radical treatments for mental illness.* New York: Basic Books; 1986. 338p.

The author, who is a neuroscientist, provides an account of the history of lobotomy and explores the dangers presented by radical therapies in all areas of medicine. Two individuals central to this history are Egas Moniz, who first performed the procedure, and Walter Freeman, who was responsible for spreading its use worldwide.

Van Nostrand's scientific encyclopedia. 6th ed. Edited by Douglas M. Consideine and Glenn D. Considine. New York: Van Nostrand Reinhold; 1983. 2 vols.

Contains over 7,300 articles, varying in length from a few lines to several pages. Larger articles contain citations to related literature. A very useful source, attractively printed.

Watson, James D. *The double helix: a personal account of the discovery of the structure of DNA.* Edited by Gunther S. Stent. New York: W.W. Norton; 1980. 298p. (Norton Critical Editions in the History of Ideas)

In addition to Watson's account of the discovery of DNA structure by himself and Francis Crick in 1953, this edition includes commentary, reviews, and original papers reporting the research. This is a highly entertaining and informative volume that describes what it is like to do scientific research, as well as explicates the breakthrough with DNA.

Wightman, William P.D. *The growth of scientific ideas.* London: Oliver and Boyd; 1951. 495p.

The development of scientific thought serves as the theme of this work, which examines certain important ideas as an avenue for study. It covers the physical and life sciences. A chronology of events is included.

Woglom, William H. *Discoverers for medicine.* New Haven, CT: Yale University Press; 1949. 229p.

Essays in this volume describe people and events surrounding thirteen advances significant to medicine. It includes chapters on such topics as blood pressure, respiration, vaccination, eyeglasses, and heredity. In addition to containing interesting illustrations, it offers a long bibliography.

Listing of References of Interest

The following citations are for reference sources which we consulted in preparing this book but did not cite. Nevertheless they are all sources worth examining.

Armytage, W. H. G. *A social history of engineering.* Cambridge, MA: MIT Press; 1961. 378p.

A very detailed history of engineering as it relates to society. Covers from ancient times to around 1960. Although the emphasis is on Great Britain, events and engineers in other countries are not neglected. Carefully written, it has good bibliographies and indexes.

Calder, Nigel. *Timescale: an atlas of the fourth dimension.* New York: Viking; 1983. 288 p.

Uses a chronological arrangement of the description of events taking place 500 million years ago up to 1982. Only eight pages cover developments since 1900, to give some idea of the emphasis on much older times. Interestingly written, with many photographs and drawings. Covers everything from dinosaurs to space ships. Has a useful index, including a historical timescale from prehistoric times to the present.

Hampel, Clifford A., ed. *The encyclopedia of the chemical elements.* New York: Reinhold; 1968. 849 p.

Devoted entirely to a description of the properties, history, and use of each of the 103 elements. Additional references are given for each article. About twenty articles on such topics as noble gases, isotopes, and origins of the elements are included.

Hodges, Henry. *Technology in the ancient world.* New York: Knopf; 1970. 297 p.

An interesting history of technology, written for the general reader. It covers the centuries from the earliest time up to the opening years of the fifth century A.D. Profusely illustrated with black and white photographs.

McGraw-Hill dictionary of science and technology. Edited by Sybil P. Parker. New York: McGraw-Hill; 1984. 942 p.

A very useful compilation of more than 35,000 terms, representing over 100 areas of science and engineering. Designed for the nonprofessional. Each definition includes an abbreviation showing the field in which the term has primary use, such as physics, mineralogy, etc.

Mind alive encyclopedia: technology. Ed. by David Chatterton. London: Chartwell Books; 1977. 240 p.

Has five main sections (electronics, astronomy, energy and engineering, technology, and industry). Describes the nature and origin of many important devices. Some of the historic background of each area discussed is included.

Modern scientists and engineers. New York: McGraw-Hill; 1980. 3 vols.

Over 1,100 biographies of outstanding scientists and engineers, each including a drawing of the person. Arranged alphabetically with detailed subject index. Also has an index by major disciplines (Scientist, Civil Engineer, etc.).

Oliver, John W. *History of American technology.* New York: Ronald Press; 1956. 676p.

A thorough treatise on the role of the United States in the development of science and engineering. It presents a detailed account of people and events from 1607 to 1955. A scholarly work, yet one which is written in a style to interest the layperson.

Parkinson, Claire L. *Breakthroughs: a chronology of great achievements in science and mathematics. 1200–1930.* Boston: G. K. Hall; 1985. 576 p.

Consists of a chronological arrangement of significant developments in science and mathematics in the years indicated. Names the persons involved and gives a brief summary of the event. Has a names, and subjects index. A useful reference tool for identifying when something was discovered and by whom.

Singer, Charles and others, eds. *History of technology.* Oxford: Clarendon Press; 1954–1984. 8 vols.

A monumental set of volumes, providing a very detailed description (often illustrated) of the history of technology and engineering. Volumes begin with prehistoric times and the closing period is around 1950. There is a separate volume for four indexes, arranged by contents of volumes, by personal names, by place names, and by subjects. This is an unusually rich source of information.

Timetable of technology. New York: Hearst Books; 1982. 240p.

Presents timetables from 1900 through 1982 for inventions and developments, covering four broad categories: communication and information; transportation and warfare; energy and industry; and medicine and food products. One set of pages is devoted to each year. There are also two dozen or so articles on recent developments in such fields as computers and communication. A useful reference tool.

Williams, Trevor I. *A short history of twentieth century technology: c.1900-c.1950.* New York: Oxford University Press; 1982. 411p.

A condensed version of later volumes in the History of Technology set. Carefully written, emphasizing highlights of the period covered.

World who's who in science: a biographical dictionary of notable scientists from antiquity to the present. Chicago: Marquis; 1968. 1855 p.

Gives brief sketches of approximately 30,000 scientists, nearly half being historical. Gives usual biographical data including fields in which involved. Contains much data on persons not easy to identify, as well as famous scientists.

Personal Name Index

Chronological Index

This is an index to *all* dates mentioned in the main entry for a particular discovery, invention, or development. Thus many discoveries and inventions appear more than once. They represent not only the year a major discovery took place but also the years in which related developments and refinements took place. By turning to the main entry the reader can readily determine what event associated with the subject of that entry took place in the year in question.

Geographical Index

Estonia

Europe

Finland

France

Germany

Greece

Holland

Hungary

India

Iran

Iraq

Ireland

Israel

Italy

U.S.S.R. (Russia)

Wales

Field of Study Index

Chemistry

Civil Engineering

Communications

Computer Science

Control Systems

Cryogenics

Crystallography

Cytology

Dentistry

Ecology

Electrical Engineering

Electricity

Electromagnetism

Electronics

Metallurgy

Meteorology

Microbiology

Mineralogy

Mining Engineering

Mycology

Naval Architecture

Navigation

Nuclear Physics

Nucleonics

Oceanography

Ophthalmology

Optics

Ordnance

Organic Chemistry

Paleontology

Petroleum Engineering

Pharmacology

Physical Chemistry

Physics